Functional Skills
Maths

City & Guilds – Level 2

This brilliant CGP book is the best way to prepare for City & Guilds Level 2 Functional Skills Maths. It covers everything you need... and nothing you don't!

Every topic is clearly explained, along with all the non-calculator methods you'll need for the latest exam. There are practice questions throughout the book <u>and</u> a realistic practice paper at the end — all with answers included.

How to access your free Online Edition

This book includes a free Online Edition to read on your PC, Mac or tablet. You'll just need to go to **cgpbooks.co.uk/extras** and enter this code:

By the way, this code only works for one person. If somebody else has used this book before you, they might have already claimed the Online Edition.

CGP — still the best! ☺

Our sole aim here at CGP is to produce the highest quality books — carefully written, immaculately presented and dangerously close to being funny.

Then we work our socks off to get them out to you — at the cheapest possible prices.

Contents

Section Three — Handling Data

Published by CGP

Editors:
Adam Bartlett, Michael Bushell, Tom Miles

With thanks to Liam Dyer and Caley Simpson for the proofreading.

ISBN: 978 1 78908 394 1

Printed by Elanders Ltd, Newcastle upon Tyne.
Clipart from Corel®

510 FUN
3 BOOKS
CLL 24138

Numbers

Split Big Numbers into Columns and Parts

1) First, you need to know the names of all the columns.
   For example, for the number 5 739 128:

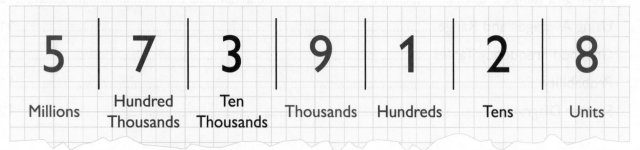

5	7	3	9	1	2	8
Millions	Hundred Thousands	Ten Thousands	Thousands	Hundreds	Tens	Units

2) You can then split any number up into parts like this:

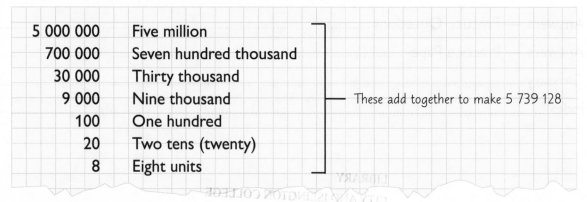

5 000 000	Five million	
700 000	Seven hundred thousand	
30 000	Thirty thousand	
9 000	Nine thousand	These add together to make 5 739 128
100	One hundred	
20	Two tens (twenty)	
8	Eight units	

Look at Big Numbers in Groups of Three

To read or write a big number, follow these steps:

1) Start from the right-hand side of the number.

2) Moving left, put a space or comma every three digits to break it up into groups of three.

3) Now going right, read each group as a separate number, as shown.

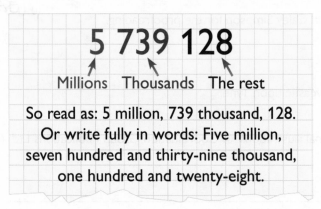

5 739 128

Millions   Thousands   The rest

So read as: 5 million, 739 thousand, 128.
Or write fully in words: Five million,
seven hundred and thirty-nine thousand,
one hundred and twenty-eight.

Negative Numbers are Less than Zero

1) A negative number is a number less than zero.

2) You write a negative number using a minus sign (–). For example, –1, –2, –3.

3) A number line is really useful for understanding negative numbers.

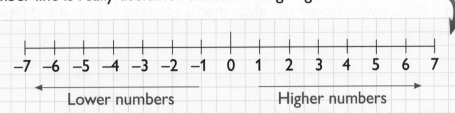

All negative numbers are to the left of zero.

All positive numbers are to the right of zero.

The further right you go, the higher the numbers get.
For example, –2 is higher than –7.

Putting Whole Numbers in Order

1) First, separate the positive and negative numbers. Then you can look at the digits.

2) Sort the numbers into groups with the same number of digits.
Negative numbers with more digits are more negative, so they're smaller.

3) Then put each group in order. Start by looking at the first digit. If the first digits are the same, move onto the second digit, then the third digit, and so on.

> **EXAMPLE:**
>
> Put these numbers in order from largest to smallest.
> 296, 1402, –64, 3, 89, 93, –45, 168, –5, 45
>
> 1) Put the numbers into groups with the same number of digits,
> separating the positive and negative numbers.
>
> | positive 4 digits | positive 3 digits | positive 2 digits | positive 1 digit | negative 1 digit | negative 2 digits |
> |---|---|---|---|---|---|
> | 1402 | 296, 168 | 89, 93, 45 | 3 | –5 | –64, –45 |
>
> 2) Put each separate group in order of size from largest to smallest.
>
> | positive 4 digits | positive 3 digits | positive 2 digits | positive 1 digit | negative 1 digit | negative 2 digits |
> |---|---|---|---|---|---|
> | 1402 | 296, 168 | 93, 89, 45 | 3 | –5 | –45, –64 |
>
> 3) So the order is:
>
> **1402, 296, 168, 93, 89, 45, 3, –5, –45, –64**

Practice Questions

1) What is 6 986 197 written in words?

   ....

2) What is seven million, two hundred and eight thousand, three hundred and fifty-two written as a number?

   ....

3) Put these lists of numbers in order from smallest to largest.

   a)   6, –8, 15, –2, 13, –16   b)   52 465, 597 246, 574 643, 14 568

   ....

   ....

4) The box below shows the temperatures recorded in six cities.

   | Moscow | –8 °C | Berlin | 4 °C | London | 6 °C |
   | Vancouver | –13 °C | Los Angeles | 19 °C | Rio | 26 °C |

   Which city has the:

   a)   highest temperature?   b)   lowest temperature?

   ....

   c)   What is the difference in temperature between Moscow and London?

   ....

5) Put this list of numbers in order from largest to smallest.
   –20 465, 957 846, –978 246, –574 643, 20 466, 957 824, –571 468

   ....

   ....

6) The number of points scored by contestants on a quiz show are shown below.
   In the quiz, you lose points for a wrong answer.

   | Lucas: –46 | Mary: –9 | Mo: 98 |
   | Oma: 154 | Torie: 151 | Joey: –54 |

   a)   Who got the lowest score?   b)   Who got the highest score?

   ....

Adding and Subtracting

You Need to Know When to Add or Subtract

1) Most of the questions you get in the test will be based on real-life situations.

2) You won't always be told whether to add or subtract (take away).

3) You'll need to work out for yourself what calculation to do.

EXAMPLE 1:

Kerry has £400. She needs to pay £110 in council tax and £246 for her electricity bill. She also wants to buy a new coat for £50.

Will Kerry have enough money left for the coat once she has paid her bills? Explain your answer.

You need to take away £110 and £246 from £400, then see how much is left.

$$4\ 0\ 0 - 1\ 1\ 0 - 2\ 4\ 6 = \boxed{44}$$

Kerry only has £44 left once she has paid her bills.
So **no**, she doesn't have enough money left for the coat.

4) You'll also need to be able to add and subtract without a calculator.

EXAMPLE 2:

Last month, an airport was used by 21 387 domestic passengers and 203 985 international passengers. How many passengers used the airport in total?

You need to add together the domestic and international passengers.

Line up the units columns and add each column from right to left.

```
  203985
+  21387
  225372
   1 1 1
```

Remember to carry when your answer is greater than 9.

So **225 372** passengers used the airport in total last month.

EXAMPLE 3:

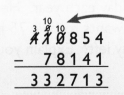

What is 410 854 − 78 141?

Line up the units columns and subtract the columns from right to left.

```
  3 10 10
  4̸ 1̸ 0̸ 8 5 4
−   7 8 1 4 1
    3 3 2 7 1 3
```

Remember, borrow from the column to the left when the top number is smaller than the bottom number.

So 410 854 − 78 141 = **332 713**

Checking Your Answer

1) Adding and subtracting are opposite calculations.

2) Once you've got your answer, you can check it using the opposite calculation.

3) You should get back to the number you started with.

EXAMPLE:

What is 92 + 25?

Answer: 92 + 25 = **117**

Check it using the opposite calculation: 117 − 25 = 92   OR   117 − 92 = 25

You only need to do one of these calculations to check your answer.

Practice Questions

1) 7643 people went to a football match in Manchester. 6391 people went to a football match in Glasgow. How many people went to these two football matches in total?

....

....

2) 25 620 people work for PFC Textiles. 9660 are under 25 years old. How many are aged 25 or over? Show how you can check your answer.

....

....

3) Zainab buys a dress for £360 and a new pair of shoes for £56. She has a voucher for £25 off her final bill. How much does Zainab have to pay after using the voucher?

....

....

4) Doug gets 190 hours of holiday each year. He used 65 hours of holiday at Christmas and 45 hours at Easter. He wants to take 75 hours for his summer holiday. Does he have enough holiday left? Explain your answer.

....

....

Multiplying and Dividing

You Need to Know When to Multiply or Divide

1) Some calculations will involve multiplication and division.

2) Like with addition and subtraction, you won't always be told whether to multiply or divide. So you'll need to work out for yourself what calculation to do.

EXAMPLE 1:

Jan needs to buy 100 large envelopes at 76p each. How much money does she need?

Answer: Each envelope costs 76p.
So you need to work out 100 times 76.

$$100 × 76 = 7600p = £76.00$$

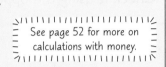
See page 52 for more on calculations with money.

3) You also need to know how to multiply and divide without a calculator for the non-calculator part of the test.

EXAMPLE 2:

A TV quiz team has twelve players. The team wins £64 320 in prize money, to be split equally between the players. How much will each player be given?

Answer: The £64 320 has to be divided between 12 players.
So you need to divide 64 320 by 12.

$$\begin{array}{r} 0\,5\,3\,6\,0 \\ 12\overline{)6\,^6 4\,^4 3\,^7 2\,0} \end{array}$$

So each player will get **£5360**.

EXAMPLE 3:

A hire car company has 15 cars. They rent each car out for £515 a week for 6 weeks. How much will the hire car company make in total?

1) First find 515 × 6.

$$\begin{array}{r} 515 \\ \times\quad 6 \\ \hline 3090 \\ \scriptstyle 3 \end{array}$$

This is the amount of money each car makes for the company.

2) Then find 3090 × 15.

$$\begin{array}{r} 3090 \\ \times\quad 15 \\ \hline 15450 \\ \scriptstyle 4 \\ +30900 \\ \hline 46350 \\ \scriptstyle 1 \end{array}$$

See page 5 for more on non-calculator addition.

So the hire car company will make **£46 350** in total.

Checking Your Answer

1) Multiplying and dividing are opposite calculations.

2) Once you've got your answer, you can check it using the opposite calculation.

3) You should get back to the number you started with.

EXAMPLE:

What is 32 × 9?

Answer: 32 × 9 = **288**

Check it using the opposite calculation: 288 ÷ 9 = 32   OR   288 ÷ 32 = 9

Practice Questions

1) What is the answer to 67 427 × 71?

....

....

....

2) What is 33 784 ÷ 6 and how much is left over?

....

....

....

3) A newsagent buys 100 chocolate bars for £40.20 and sells them on for 67p each. How many does he need to sell to make his money back?

....

....

4) Jake buys 1470 roses for £0.77 per rose. He sells bunches of 6 roses for £19.50.

a) How much did Jake spend on the roses in total?

....

b) How much profit did Jake make from selling all of the roses?

....

....

Some Questions Need Answers that are Whole Numbers

1) Real-life division questions can be tricky.
   You won't always end up with a whole number.

2) But sometimes, you'll need to give a whole number as your answer.

EXAMPLE 1:

Glenn is laying a new driveway. He needs 663 slabs for the driveway.
Slabs are available in packs of 12.
How many packs does Glenn need to buy?

Calculation: $663 \div 12 = 55.25$

He can't buy 55.25 packs, so you need to give your answer as a whole number.

55.25 is between 55 and 56. There aren't enough slabs in 55 packs.
So Glenn will have to buy **56 packs** and have 9 slabs extra.

EXAMPLE 2:

Tina uses 2.4 m² of material to make a dress.
How many dresses can she make out of 9582 m² of material?

Calculation: $9582 \div 2.4 = 3992.5$

You can't have 3992.5 dresses, so you need
to give your answer as a whole number.

3992.5 is between 3992 and 3993. There isn't enough material
for 3993 dresses. So Tina will only be able to make **3992 dresses**.

Practice Questions

1) A large egg box holds 12 eggs. A farm has 1450 eggs.
   How many large egg boxes do they need to carry all of the eggs?

   ....

2) A school has 47 classes of 26 students each. A box of pencils contains 24 pencils. How
   many boxes of pencils does the school need to buy so that all of the students have a pencil?

   ....

   ....

   ....

Order of Operations

Calculations with Several Steps

1) You'll sometimes need to do calculations that have several steps.

2) You need to be careful about the order that you do a calculation.
   If you do it in the wrong order, you might get the wrong answer.

Order of Calculations (BIDMAS)

1) BIDMAS tells you the order in which calculations should be done:

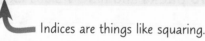

$\underline{B}$rackets, $\underline{I}$ndices, $\underline{D}$ivision, $\underline{M}$ultiplication, $\underline{A}$ddition, $\underline{S}$ubtraction

Indices are things like squaring.

2) Work out brackets first, then indices, then multiply and divide,
   then add and subtract.

3) For multiplying and dividing, you work from left to right.
   You also work from left to right when adding and subtracting.

EXAMPLE 1:

What is 800 + 2400 ÷ 400?

1) Follow BIDMAS —
   do the division first:
   $800 + 2400 \div 400 = 800 + 6$

2) Then do the addition:
   $800 + 6 = \mathbf{806}$

If you didn't follow BIDMAS, you'd do
800 + 2400 = 3200, and
3200 ÷ 400 = 8 (which is wrong).

EXAMPLE 2:

What is $(5 \times 4 + 10)^2$?

1) Do the calculation in the brackets
   first. Use BIDMAS for the
   calculation in the brackets and do
   the multiplication first:
   $5 \times 4 = 20$
   $20 + 10 = 30$

2) Do the indices: $30^2 = 30 \times 30 = 900$.
   So $(5 \times 4 + 10)^2 = \mathbf{900}$

EXAMPLE 3:

What is $\dfrac{5 + 15}{2^2}$?

Answer: Work out the top and bottom parts first.
Then divide the top by the bottom.

$$\frac{5 + 15}{2^2} = \frac{(5 + 15)}{(2^2)} = \frac{20}{4} = 20 \div 4 = \mathbf{5}$$

Imagine that there are
invisible brackets on
the top and on the
bottom of the fraction.

Order of Calculations Using a Calculator

1) Calculators follow BIDMAS.

2) Be careful when typing a calculation into your calculator. If you get it in the wrong order, your calculator will give the wrong answer.

3) If your calculator has bracket buttons, make sure you use them. The calculator will work out the bits inside the brackets before it does the rest of the calculation.

4) If your calculator doesn't have bracket buttons, just work out the bits in brackets for yourself first, then enter the result into your calculator.

EXAMPLE:

What is $8^2 \div (24 + 8)$?

1) Work out the bit in the brackets first: 　2　4　+　8　= 　　32

2) Next work out the indices: 　8　x^2　= 　　64

3) Put these two answers into the calculation:

　6　4　÷　3　2　= 　　2

Practice Questions

1) Calculate:

a) $(2 + 6)^2 - 3$

b) $2 + (6^2 - 3)$

2) What is $(360 + 60) \div (7 \times 3)$?

3) What is $5^2 - 4 \times (36 \div 6)$?

4) What is $\dfrac{16 \times 4 - 14}{3^2 - 4}$?

Fractions

Fractions Show Parts of Things

1) If something is divided up into equal parts, you can show it as a fraction.

2) There are two bits to every fraction:

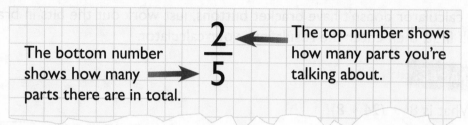

The bottom number shows how many parts there are in total. → $\dfrac{2}{5}$ ← The top number shows how many parts you're talking about.

3) An improper fraction is a fraction which has a top number that is bigger than the bottom number. For example $\dfrac{6}{5}$ or $\dfrac{14}{3}$. They're greater than the number 1.

4) You treat improper fractions in just the same way as normal fractions.

Equivalent Fractions

1) Equivalent fractions are equal in size, but the numbers on the top and bottom are different.

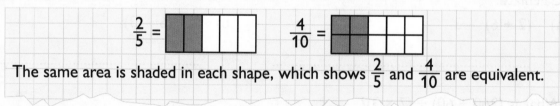

$\dfrac{2}{5} =$ $\dfrac{4}{10} =$

The same area is shaded in each shape, which shows $\dfrac{2}{5}$ and $\dfrac{4}{10}$ are equivalent.

2) To get from one fraction to an equivalent one, multiply or divide the top AND the bottom by the same number.

EXAMPLE:

What fraction is equivalent to $\dfrac{4}{9}$ but has 27 on the bottom?

Answer: To get from 9 to 27, you multiply by 3. So to find the new top number, you need to multiply the top by 3.

new top number = 4 × 3 = 12, so $\dfrac{4}{9} \overset{\times 3}{\underset{\times 3}{=}} \dfrac{12}{27}$.

Writing Fractions in their Simplest Form

1) If the top and bottom of the fraction are both multiples of the same number, then the fraction can be simplified using equivalent fractions.

2) A fraction is in its simplest form if the numbers on the top and the bottom are as small as possible — they can't be divided by the same number to give whole numbers.

EXAMPLE 1:

Write the fraction $\frac{10}{15}$ in its simplest form.

Both 10 and 15 are divisible by 5, so divide the top and bottom of the fraction by 5.

$$\frac{10}{15} \overset{\div 5}{\underset{\div 5}{=}} \frac{2}{3}$$

EXAMPLE 2:

Write the fraction $\frac{36}{144}$ in its simplest form.

1) Both 36 and 144 are divisible by 12, so divide the top and bottom of the fraction by 12.

$$\frac{36}{144} \overset{\div 12}{\underset{\div 12}{=}} \frac{3}{12}$$

2) Then 3 and 12 can both be divided by 3, so the fraction can be simplified further.

$$\frac{3}{12} \overset{\div 3}{\underset{\div 3}{=}} \frac{1}{4}$$

Expressing One Number as a Fraction of Another

1) To write one number as a fraction of another, you write the first number as the top part of the fraction and the second number as the bottom part.

2) You'll then need to simplify the fraction (see above).

EXAMPLE:

What is 30 as a fraction of 75? Write your answer in its simplest form.

1) Write the first number on top and the second number on the bottom. ⟶ $\frac{30}{75}$

2) Both 30 and 75 can be divided by 5, so the fraction can be simplified.

$$\frac{30}{75} \overset{\div 5}{\underset{\div 5}{=}} \frac{6}{15}$$

3) 6 and 15 can both be divided by 3, so the fraction can be simplified further.

$$\frac{6}{15} \overset{\div 3}{\underset{\div 3}{=}} \frac{2}{5}$$

Practice Questions

1) 13 people take their driving test on the same day. 9 of them pass.
   What fraction of people didn't pass?

....

2) What number should go in the box so that $\frac{\square}{81}$ is equivalent to $\frac{5}{9}$?

....

3) What number should go in the box so that $\frac{\square}{3}$ is equivalent to $\frac{28}{12}$?

....

4) Write these fractions in their simplest form.

   a) $\frac{7}{42}$ b) $\frac{75}{150}$ c) $\frac{56}{72}$ d) $\frac{324}{72}$

   ....

   ....

5) Write these numbers as a fraction of 120. Simplify your answers.

   a) 24   b) 75   c) 90   d) 192

   ....

   ....

6) In the month before a show, Mai's music teacher recommended that she should
   spend 68 hours practising. Mai actually practised for 60 hours in the month
   before the show. Write how long Mai actually practised as a fraction of what
   her teacher recommended. Give your answer in its simplest form.

....

....

7) Last year, Jeremy had two jobs. As a waiter, he earned £90 per weekend for 26 weekends.
   As a climbing instructor, he earned £720 a week for 13 weeks. What fraction of his total
   income was from being a waiter? Give your answer in its simplest form.

....

....

Mixed Numbers

1) Mixed numbers are when you have a whole number part and a fraction part together.

$1\frac{1}{4}$ $2\frac{3}{5}$

one and one quarter   two and three fifths

2) To convert a mixed number into a fraction, you first need to find the new top number.
Multiply the whole number part by the bottom number of the fraction part.
Add this to the top number of the fraction to get the new top number.

3) The bottom number stays the same.

EXAMPLE 1:

What fraction is equal to $4\frac{7}{12}$?

1) Find the new top number. Multiply the whole number part by the bottom number of the fraction. Then add this to the top number.

$4 \times 12 = 48$, so the new top number is $48 + 7 = 55$

2) The bottom number stays the same, so $4\frac{7}{12} = \frac{55}{12}$.

4) To convert from a fraction to a mixed number, you need to divide the top number by the bottom number. You should do this without a calculator so you get a remainder.

5) The whole number part of the mixed number will be the result of the division.
The fraction part will have the remainder on the top. The bottom number stays the same.

EXAMPLE 2:

Convert $\frac{25}{8}$ into a mixed number.

1) Do the division: $25 \div 8 = 3$ remainder 1

2) The whole number is 3 and the top part of the fraction is 1, so the mixed number is $3\frac{1}{8}$.

Practice Questions

1) Convert the following mixed numbers into fractions.

   a) $2\frac{3}{5}$ b) $1\frac{3}{7}$ c) $3\frac{5}{12}$ d) $2\frac{2}{13}$

   ....

2) Convert the following fractions into mixed numbers.

   a) $\frac{7}{3}$ b) $\frac{43}{4}$ c) $\frac{21}{5}$ d) $\frac{27}{8}$

   ....

Comparing Fractions

1) Fractions are just numbers, so they can be put in order of size.

2) Look at the bottom numbers. If they're the same, then you can just order the fractions by looking at the top numbers.

3) If the bottom numbers are different, you'll need to use equivalent fractions. Turn the fractions into equivalent fractions with the same bottom number.

4) Then you can use the top numbers to order them.

5) If you have to compare mixed numbers and improper fractions, convert the mixed numbers into fractions first.

EXAMPLE:

Which is bigger, $\frac{5}{4}$ or $1\frac{3}{7}$?

1) Change the mixed number into a fraction.

$1 \times 7 = 7$, so the new top number $= 7 + 3 = 10$. So $1\frac{3}{7} = \frac{10}{7}$

2) The bottom numbers are different so you need to find equivalent fractions.

3) $4 \times 7 = 28$, so multiply the top and bottom of each fraction to get 28 on the bottom.

$$\overset{\times 7}{\frac{5}{4}} = \frac{35}{28} \qquad \overset{\times 4}{\frac{10}{7}} = \frac{40}{28}$$
$$\underset{\times 7}{} \qquad \underset{\times 4}{}$$

4) Now compare the top numbers.

40 is bigger than 35, so $1\frac{3}{7}$ is bigger than $\frac{5}{4}$.

Practice Questions

1) Which fraction is larger, $\frac{3}{4}$ or $\frac{30}{36}$?

....

2) Put the following fractions in order from smallest to largest.

a) $\frac{120}{400}, \frac{2}{5}, \frac{5}{20}$ b) $\frac{3}{20}, \frac{6}{24}, \frac{9}{54}$

....

....

3) Mike and Leo are running laps of a field. Mike has run $5\frac{7}{10}$ laps and Leo has run $\frac{133}{25}$ laps. Who has run the furthest?

....

....

Adding and Subtracting Fractions

1) You can only add and subtract fractions if the bottom numbers are the same.

2) If the bottom numbers are not the same, you will need to use equivalent fractions. Turn the fractions into equivalent fractions with the same number on the bottom.

3) Then you just add or subtract the top numbers. The bottom number stays the same.

EXAMPLE:

What is $\frac{3}{10} + \frac{3}{5} - \frac{1}{2}$? Write your answer in its simplest form.

1) The bottom numbers are not the same — use equivalent fractions to find fractions with the same bottom number.

$$5 \times 2 = 10, \text{ so } \frac{3}{5} = \frac{6}{10} \quad \text{(×2)} \qquad 2 \times 5 = 10, \text{ so } \frac{1}{2} = \frac{5}{10} \quad \text{(×5)}$$

2) Add and subtract the top numbers. The bottom number stays the same.

$$\frac{3}{10} + \frac{6}{10} - \frac{5}{10} = \frac{3+6-5}{10} = \frac{4}{10}$$

3) Convert your answer into its simplest form.

Both 4 and 10 are divisible by 2, so $\frac{4}{10} = \frac{2}{5}$ (÷2).

Addition and Subtraction Involving Mixed Numbers

1) To add or subtract mixed numbers, first split up the whole number and fraction parts.

2) Add or subtract the two parts separately and then add them back together.

3) If the fraction part calculation gives an improper fraction, you'll need to change it to a mixed number and add the whole number part to the others.

EXAMPLE 1:

What is $2\frac{3}{4} + 1\frac{3}{8}$?

1) Split up the whole number and fraction parts:

$$2\frac{3}{4} + 1\frac{1}{8} = 2 + 1 + \frac{3}{4} + \frac{3}{8}$$

2) Add up each part:

$$2 + 1 = 3 \qquad \qquad \frac{3}{4} + \frac{3}{8} = \frac{6}{8} + \frac{3}{8} = \frac{9}{8} = 1\frac{1}{8}$$

See page 15 for converting between improper fractions and mixed numbers.

3) Put the two parts back together:

$$3 + 1\frac{1}{8} = 4\frac{1}{8}$$

EXAMPLE 2:

What is $4\frac{2}{5} - 2\frac{1}{10}$?

1) Split the whole number and fraction parts:

$$4\frac{2}{5} - 2\frac{1}{10} = 4 - 2 + \frac{2}{5} - \frac{1}{10}$$

2) Subtract each part:

$$4 - 2 = 2 \qquad\qquad \frac{2}{5} - \frac{1}{10} = \frac{4}{10} - \frac{1}{10} = \frac{3}{10}$$

3) Put the two parts back together:

$$2 + \frac{3}{10} = 2\frac{3}{10}$$

Take care when subtracting mixed numbers. You need to subtract the different parts — but when you're putting the two parts back together, you need to add them.

Practice Questions

1) What is $\frac{1}{4} + \frac{2}{5}$?

....

....

2) What is $\frac{4}{5} - \frac{3}{10}$?

....

....

3) What is $\frac{5}{3} + \frac{3}{7}$?

....

....

4) What is $\frac{17}{8} - \frac{3}{5}$?

....

....

5) What is $1\frac{2}{3} + 2\frac{5}{6}$?

....

....

6) What is $4\frac{5}{8} - 2\frac{1}{4}$?

....

....

7) Jasper is going on holiday. The airline says he can take a maximum of 22 kg of luggage onto the plane in total. His main suitcase weighs $18\frac{3}{4}$ kg, his carry-on bag weighs $2\frac{3}{8}$ kg and he also has another small bag that weighs $\frac{4}{5}$ kg.
Will he be allowed to take all of his bags onto the plane? Explain your answer.

....

....

Using Fractions With a Calculator

1) $\frac{3}{4}$ is just another way of writing $3 \div 4$.

2) If your calculator doesn't have a fraction button, you can still use it to turn fractions into decimals by dividing. See page 35 for turning fractions into decimals.

3) Once you've finished any calculations, don't forget to give your answer as a fraction again. See page 36 for turning decimals back into fractions.

4) Finally, make sure to put the fraction in its simplest form.

EXAMPLE:

Calculate $\frac{201}{268} - \frac{28}{40}$.

1) Convert the fractions into decimals.

$201 \div 268 = 0.75$ $\qquad$ $28 \div 40 = 0.7$

2) Then do the calculation.

$0.75 - 0.7 = 0.05$

3) Finally convert the decimal back to a fraction and simplify it.

$0.05 = \frac{5}{100}$. Both 5 and 100 are divisible by 5, so $\frac{5}{100} \overset{\div 5}{\underset{\div 5}{=}} \frac{1}{20}$.

Practice Questions

1) Calculate $\frac{5}{2} + \frac{15}{4} - \frac{21}{8}$. Give your answer in its simplest form.

....

....

2) Calculate $\frac{5}{2} - \frac{2}{16} + \frac{1}{4}$. Give your answer in its simplest form.

....

....

3) Anoush and Martin complete a quiz. Anoush gets $\frac{15}{24}$ of the questions correct and Martin gets $\frac{84}{96}$ right. What is the difference in the fraction of the questions they got correct?

....

....

Finding Fractions 'Of' Something

1) You might need to calculate the 'fraction of' something.

2) In these cases, 'of' means 'times' (multiply).

EXAMPLE 1:

What is $\frac{1}{32}$ of 14 656?

1) 'Of' means 'times' (×), so $\frac{1}{32}$ of 14 656 is the same as $\frac{1}{32}$ × 14 656.

2) If your calculator doesn't have a fraction button, you can type

fractions in by dividing the top by the bottom: $\frac{1}{32}$ = 1 ÷ 32

3) So the full calculation you need to do is: 1 ÷ 32 × 14 656 = **458**

EXAMPLE 2:

A survey asks 32 820 people whether they are satisfied with their job. $\frac{7}{12}$ of the people asked say 'no'. How many people say no?

You need to calculate $\frac{7}{12}$ of 32 820.

1) 'Of' means 'times' (×), so $\frac{7}{12}$ of 32 820 is the same as $\frac{7}{12}$ × 32 820.

2) 7 ÷ 12 × 32 820 = 19 145   So **19 145** people said no.

Practice Questions

1) What is $\frac{12}{50}$ of 40 850? ....

2) What is $\frac{2}{11}$ of 63 162? ....

3) A hair salon has 840 clients booked in for the month.

a) Three-eighths of the clients cancel their appointments. How many clients cancel?

....

b) Of the remaining clients, seven-fifteenths are booked in for a cut and colour. How many clients are booked in for a cut and colour?

....

....

Decimals

Not All Numbers Are Whole Numbers

1) Decimals are numbers with a decimal point (.) in them. For example, 0.5, 1.3.

2) They're used to show the numbers in between whole numbers.

> The number 32.95 is a bit smaller than the number 33.
>
> The number 32.5 is exactly halfway between the numbers 32 and 33.
>
> The number 32.257 is slightly bigger than 32.25.

3) The first digit after a decimal point shows tenths, the second digit shows hundredths, and the third digit shows thousandths.

4) You can show decimals on a number line:

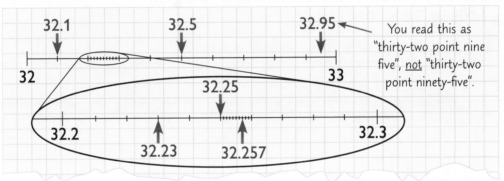

You read this as "thirty-two point nine five", _not_ "thirty-two point ninety-five".

How to Put Decimals in Order

You might need to arrange a list of decimal numbers in order of size.

EXAMPLE:

Put these decimals in order from smallest to largest: 1.02, 0.2004, 0.204, 0.002.

1) Put the numbers into a column, lining up the decimal points.

2) Make all the numbers the same length by filling in extra zeros at the ends.

3) Look at the numbers before the decimal point.
   Arrange the numbers from smallest to largest.

4) If any of the numbers are the same, move onto the numbers after the decimal point. Arrange the numbers from smallest to largest.

Step 1:	Step 2:	Step 3:	Step 4:
1.02	**1.0200**	**0.2004**	**0.0020**
0.2004	0.2004	0.2040	0.2004
0.204	0.2040	0.0020	0.2040
0.002	0.0020	1.0200	1.0200

The order is: **0.002, 0.2004, 0.204, 1.02.**

Practice Questions

1) Write 55.354 in words.

....

2) Which number is larger?

    a)   6.248 or 62.48   b)   34.658 or 34.648   c)   46.83 or 46.03

    ....

3) Which one of the following lists is in increasing order? Tick your answer.

    ☐   0.185, 0.545, 0.0084   ☐   0.545, 0.185, 0.0084

    ☐   0.0084, 0.185, 0.545   ☐   0.0084, 0.545, 0.185

4) Put these weights in order of size from smallest to largest:
0.81 kg, 1.008 kg, 0.8 kg, 0.801 kg

....

....

Adding Decimals

1) Write the numbers one above the other, lining up the decimal points.

2) Make sure all the numbers have the same number of decimal places by filling in extra zeros at the ends where needed.

3) Write a decimal point on your answer line, lined up with the decimal points in the numbers above.

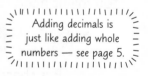

Adding decimals is just like adding whole numbers — see page 5.

4) Add up the columns from right to left. Remember to carry digits to the next column when you get a total that is greater than 9.

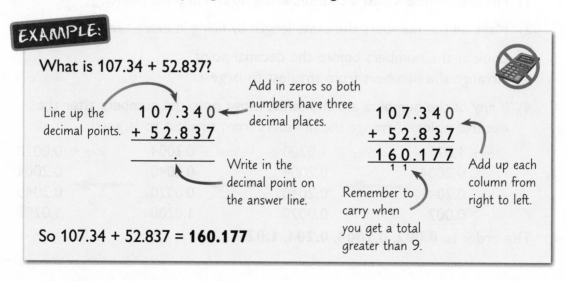

EXAMPLE:

What is 107.34 + 52.837?

Line up the decimal points.

Add in zeros so both numbers have three decimal places.

```
  1 0 7.3 4 0
+   5 2.8 3 7
```

Write in the decimal point on the answer line.

```
  1 0 7.3 4 0
+   5 2.8 3 7
  1 6 0.1 7 7
      1 1
```

Add up each column from right to left.

Remember to carry when you get a total greater than 9.

So 107.34 + 52.837 = **160.177**

Subtracting Decimals

1) Write the numbers one above the other, with the biggest number on top, lining up the decimal points.

2) Make sure all the numbers have the same number of decimal places by filling in extra zeros at the ends where needed.

3) Write a decimal point on your answer line, lined up with the decimal points in the numbers above.

> Subtracting decimals is just like subtracting whole numbers — see page 5.

4) Subtract the numbers in each column from right to left. Remember to borrow when the top digit is smaller than the bottom one.

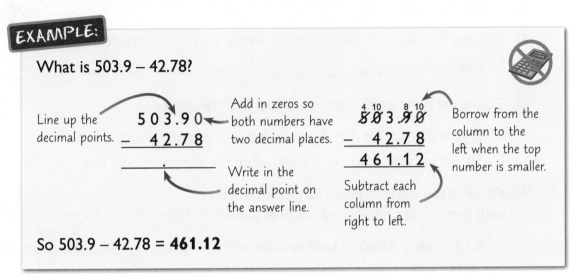

EXAMPLE:

What is 503.9 − 42.78?

Line up the decimal points.

Add in zeros so both numbers have two decimal places.

Write in the decimal point on the answer line.

$$503.90$$
$$-\ \ 42.78$$

$$\overset{4\ \ 10\ \ \ \ \ \ 8\ \ 10}{\cancel{5}\cancel{0}3.\cancel{9}\cancel{0}}$$
$$-\ \ 42.78$$
$$\overline{461.12}$$

Borrow from the column to the left when the top number is smaller.

Subtract each column from right to left.

So 503.9 − 42.78 = **461.12**

Practice Questions

1) What is 88.464 + 75.8?

....

....

....

2) What is 451.64 − 85.6?

....

....

....

3) Tom has £262.98 in his bank account. He pays in a cheque for £56.23, then spends £39.47 on petrol and £41.42 in the supermarket. How much money is left in his account?

....

....

....

Multiplying Decimals

1) Start by ignoring the decimal points. Just do the multiplication with whole numbers.

2) Count the total number of digits after the decimal points in the original numbers.

3) Write in the decimal point to make the answer have the same number of decimal places. For example, 1.2 has **1** digit after the decimal point and 2.45 has **2** digits after the decimal point, so 1.2 × 2.45 would have 1 + 2 = 3 digits after the decimal point.

EXAMPLE:

What is 5.15 × 0.6?

1) Ignoring the decimal points, do the whole number multiplication.

515 × 6 = 3090 ◄── *This multiplication is shown on page 7.*

2) Count the number of digits after the decimal points.

5.$\underset{2}{15}$ × 0.$\underset{1}{6}$ has 2 + 1 = 3 digits after the decimal point.
So the answer will have 3 digits after the decimal point too.

3) Write in the decimal point to give a number with the correct number of decimal places:

5.15 × 0.6 = 3.$\underset{3}{090}$ = **3.09** ◄── *You can leave out any zeros at the end if you're not told how many places to give your decimal answer to.*

Dividing a Decimal by a Whole Number

1) When dividing a decimal by a whole number, set the question up as you would for whole numbers.

2) Write a decimal point on the answer line directly above the decimal point in the number you're dividing.

3) Then do the division using the same method as for whole numbers.

EXAMPLE:

An 8-storey building is 37.6 m tall. Each floor is the same height. How high is each floor?

Answer: You need to divide 37.6 by 8.

1) Write the division with the number you're dividing by on the left.

$8\overline{)37.6}$

2) Put a decimal point right above the one in the number you're dividing.

$8\overline{)37.6}^{\,\cdot}$

3) Use the same method to do the division as for whole numbers.

$8\overline{)3^37^56}$ → 04.7

So each floor is **4.7 m** high.

Dividing Any Number by a Decimal

1) Write the division as a fraction. The number you're dividing by goes on the bottom.

2) Multiply the top and bottom by 10, 100 or 1000 to remove
   any decimals and leave whole numbers on the top and bottom.
   You must multiply the top and the bottom by the same thing.

3) Do the whole number division to give the answer.

EXAMPLE:

What is $45.18 \div 0.03$?

1) Write the division as a fraction.

$$45.18 \div 0.03 = \frac{45.18}{0.03}$$

2) Multiply the top and bottom by 100 to get an equivalent fraction:

$$\frac{45.18}{0.03} = \frac{4518}{3}$$

3) Do the whole number division $4518 \div 3$.

$$3 \,\overline{)\, 4^15\,1^18}$$
$$\quad\; 1\,5\,0\,6$$

Don't divide by 100
at the end — the
final answer is 1506.

So $45.18 \div 0.03 = \mathbf{1506}$

Practice Questions

1) Calculate:

a) 3.48×1.6

b) 82.3×5.9

....

....

....

....

c) $41.08 \div 4$

d) $78.4 \div 0.8$

....

....

2) Kaiko works 37.5 hours a week. He needs to divide this time equally between
   three different projects. How many hours should he spend on each project?

....

3) An office has 8 people in it. They have a total of £20.40 to spend on lunch
   for the whole office. How much do they have to spend on lunch per person?

....

Rounding and Estimating

Rounding off Decimals

You can sometimes get an answer with lots of numbers after the decimal point.
Instead of writing down the whole thing, you can shorten the answer by rounding.

> 1) To round a number to a given number of decimal places, first identify
>   the position of the 'last digit'. For example, if you're rounding to
>   2 decimal places, it's the second digit after the decimal point.
>
> 2) Look at the digit to the right of this — it's called the 'decider'.
>
> 3) If the decider is 5 or more, round up the last digit by one.
>   If the decider is 4 or less, leave the last digit as it is.
>
> 4) Don't write down any digits after the last digit (even any 0s).

If the last digit is a 9 and you have to round it up, you would need to round 9 to 10.
This means the last digit becomes 0 and you add 1 to the digit to the left of the last digit.

EXAMPLES:

1) Round 2.842 to two decimal places.

   You want two digits after the decimal point, so the decider
   is the third digit after the decimal point.

   Last digit   2.8 4 2   Decider
               1 2 3

   The decider is 2, which is less than 5, so leave the last digit as it is.
   So the answer is **2.84**.

2) Round 5.398 to two decimal places.

   Last
   digit   4 0
       5.398 → 5.39 → **5.40**   The question asks for two decimal
                                      places, so you need to include the
       Decider — round up   0 at the end — it's not just 5.4.

Rounding to the Nearest Whole Number

You can round to the nearest whole number in a similar way.
This time, the decider is the first digit after the decimal point.

EXAMPLE:

Round 25.65 to the nearest whole number.

The decider is the first digit after the decimal point, so it's 6.
This is more than 5, so the answer is **26**.

Use Rounding to Estimate the Answers to Calculations

1) You can check your answers in tricky calculations by estimating.

2) Round each number in the calculation so that all the digits apart from the first one are zero (for example to the nearest whole number or the nearest 10). Then do the calculation using the rounded numbers.

3) If your estimate is close to the answer you got, then your answer is probably correct.

 **EXAMPLE:**

Vishaal earns £3.80 an hour for a paper round.
One day, his round takes him 2.25 hours and he gets £5 in tips.
Estimate how much Vishaal made from his paper round that day.

1) Round each number so that all the digits apart from the first one are zero.

£3.80 will round to £4   2.25 rounds to 2   £5 stays as £5

2) Then do the calculation: £4 × 2 + £5 = £13

Vishaal made about **£13** from his paper round.

Practice Questions

1) Round these numbers to the given number of decimal places.

a)   1.024 to 2 decimal places

b)   3.57896 to 1 decimal place

....

....

2) Round each of these numbers to the nearest whole number.

a)   51.84

b)   5.47

....

....

3) A race track is made of 3 sections. Daniel does a lap which takes 18.638 secs in section 1, 33.392 secs in section 2 and 18.779 secs in section 3. Estimate how long Daniel's lap took.

....

....

4) The formula for the area of a circle is $A = \pi r^2$ where A is the area and r is the radius. Estimate the area of a circle where $\pi = 3.142$ and $r = 8.6$ cm.

....

....

Percentages

Understanding Percentages

1) 'Per cent' means 'out of 100'. % is a short way of writing 'per cent'.

2) 20% means twenty per cent. This is the same as 20 out of 100. 100% represents the whole amount of something.

3) You can write any percentage as a fraction or a decimal.

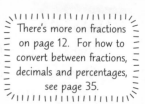

There's more on fractions on page 12. For how to convert between fractions, decimals and percentages, see page 35.

$$20\% = \frac{20}{100} = 0.2$$

Calculating the 'Percentage Of' Something

1) If you need to calculate the 'percentage of' something, multiply by the decimal that's equivalent to the percentage.

EXAMPLE 1:

What is 17% of 1580?

1) Turn the percentage into a decimal: 17% = 17 ÷ 100 = 0.17

2) Multiply by the decimal: 0.17 × 1580 = **268.6**

2) You can work out a percentage without a calculator by breaking it down into parts which are easier to work out.

EXAMPLE 2:

A hotel has 500 rooms. 63% of the rooms have been booked. How many rooms have been booked?

1) Break down the percentage into easier parts:
$$63\% = 50\% + 10\% + (1\% + 1\% + 1\%)$$

2) Work out the parts: 50% = 500 ÷ 2 = 250
10% = 500 ÷ 10 = 50
1% = 500 ÷ 100 = 5

3) Add up to find 63%:
$$63\% = 50\% + 10\% + (1\% + 1\% + 1\%)$$
$$= 250 + 50 + (5 + 5 + 5)$$
$$= 315$$

So **315** rooms have been booked.

Practice Questions

1) What is 21% of 700?

....

2) What is 72% of 275?

....

....

3) What is 16% of 15 670?

....

4) A flight from London to Madrid has 560 passengers. 37.5% of the passengers are vegetarian. How many passengers are vegetarian?

....

5) Asif is a dentist. He sees 3900 patients in a year. Of these, 28% need to have a filling. How many patients need a filling?

....

Expressing a Number as a Percentage of Another

1) If you need to express one number as a percentage of another, then you need to divide.

2) Divide the first number by the second number to get a decimal.

3) Then convert that decimal into a percentage by multiplying by 100.

EXAMPLE 1:

What is 56 as a percentage of 160?

1) Divide the first number by the second:

$$56 \div 160 = 0.35$$

2) Convert the result into a percentage by multiplying by 100:

$$0.35 \times 100 = \textbf{35\%}$$

EXAMPLE 2:

Kevin sleeps for 7.5 hours each day. There are 24 hours in a day.
What percentage of the day is Kevin asleep?

Answer: Divide the number of hours Kevin sleeps for by the total hours
in a day. Then convert the answer into a percentage.

1) $7.5 \div 24 = 0.3125$ 2) $0.3125 \times 100 =$ **31.25%**

Practice Questions

1) What is 52 as a percentage of 200?

....

2) What is 1825 as a percentage of 2920?

....

3) What is 32 616 as a percentage of 86 976?

....

4) A hotel has 45 rooms in total. 36 of them are being used by guests.
   What percentage of the rooms are being used?

....

....

5) Maggie has used 36 of her 80 hours of holiday for the year.
   What percentage of her holiday hours does she have left?

....

....

6) Brian and Liv own ice cream shops. One day, Brian has 88 customers.
   66 of them have sprinkles. Liv has 120 customers and 80 of them have sprinkles.
   In whose shop did a higher percentage of customers have sprinkles? Explain your answer.

....

....

Percentage Change

Calculating a Percentage Increase

1) Sometimes, you might need to calculate a percentage increase.

2) To do this, add the percentage increase to 100% and change the answer to a decimal.

3) Multiply by this decimal to find the increased value.

EXAMPLE:

In 2016, a town's population was 2500. By 2019, it had increased by 6%. How many people live in the town in 2019?

1) Add 6% to 100% and convert to a decimal:
   6% + 100% = 106% = 106 ÷ 100 = 1.06

2) Multiply the original value by 1.06 to find the increased value:
   2500 × 1.06 = 2650

So **2650** people live in the town in 2019.

You could also do this by finding 6% of 2500 and then adding it on.

Practice Questions

1) Jim earns £21 000 a year. He's given a pay rise of 2%.
   How much will Jim earn after his pay rise?

   ....

   ....

2) A company made a profit of £35 000 last year. This year, its profits have increased by 5%.
   How much profit did the company make this year?

   ....

   ....

3) Last year, a florist took orders from 26 000 customers. This year, orders increased by 75.2%.
   How many customers placed an order this year?

   ....

   ....

Calculating a Percentage Decrease

1) You might also need to find a percentage decrease.

2) To find a percentage decrease, you need to take away the percentage from 100%.

3) Convert the new percentage into a decimal and multiply to find the decreased value.

EXAMPLE:

Last year, a company made a profit of £20 000. This year, profits are down by 14.5%. How much profit will the company make this year?

1) Take 14.5% away from 100% and convert into a decimal:

$$100\% - 14.5\% = 85.5\% = 85.5 \div 100 = 0.855$$

2) Multiply the original value by 0.855 to find the decreased value:

$$£20\,000 \times 0.855 = £17\,100$$

You could also do this by finding 14.5% of 20 000 and then taking it away.

The company will make **£17 100** profit.

Practice Questions

1) After the January sales, a retail outlet reduces the number of items with a discount on them by 40%. During the sale, 120 products had a discount on them. How many products have a discount on them now?

....

....

2) Last year, Isla ran a 10 km race in 60 minutes. She completed it 5% faster this year. How long did it take her to run the 10 km this year?

....

....

3) Last month, Dave used 16 GB of data on his mobile phone. This month, he used 31.25% less. How much data did he use this month?

....

....

Finding the Original Value

1) You could be given the result of a percentage change and be asked to work backwards to find the original value.

2) First, write the amount that you're given as a percentage of the original value.

3) Divide to find 1% of the original value.

4) Multiply by 100 to give the original value (= 100%).

An antique vase increases in value by 15% to £3450.
Find what it was worth before the rise.

1) The value has increased by 15%, so £3450
is 100% + 15% = 115% of the original value: 115% = £3450

2) Divide by 115 to find 1%:   1% = £30

3) Multiply by 100 to find 100%:   100% = **£3000**

5) If the percentage change is a reduction, then the percentage of the original value will be less than 100%. For example, if a value is reduced by 10% then the amount you're given is 100% − 10% = 90% of the original value.

Practice Questions

1) Alistair rents a new flat which is 15% bigger than his old flat. His new flat has a floor area of 80.5 m². What was the floor area of Alistair's old flat?

....

....

2) The local gym has 135 members this month. This is an 8% increase on last month. How many members did it have last month?

....

....

3) Pia has paid off 45% of a loan and now only has £1980 left to pay. How much was the original loan?

....

....

Finding the Percentage Change

1) Sometimes, you'll be given the result after a percentage change and the original value, and you'll need to work out what the percentage increase or decrease was.

2) To do this, divide the new value by the original value to give a decimal.

3) Convert this decimal into a percentage and then subtract 100% to get the percentage change.

4) If the percentage is negative, the change was a decrease.

EXAMPLE:

When Anoush bought her car, it was worth £32 000.
The car is sold 5 years later for £24 000.
What is the percentage change in the value of the car?

1) First, divide the new value by the original value:

$$24\,000 \div 32\,000 = 0.75$$

2) Convert the decimal to a percentage: $0.75 \times 100 = 75\%$

3) Subtract 100% to find the change: $75\% - 100\% = -25\%$

So the value of the car has decreased by **25%**.

Practice Questions

1) 5 years ago, 3000 turtles nested on a beach. This year, 3090 nested on the beach.
   What is the percentage change in the number of turtles nesting on the beach?

   ....

2) Majid's car had a mileage of 12 500 miles at the start of the year. It is now 14 750 miles.
   What is the percentage change in the mileage of Majid's car?

   ....

   ....

3) The turnout at a polling station 5 years ago was 3725. At a recent election,
   the turnout was 3576. Work out the percentage change in voter turnout.

   ....

   ....

Fractions, Decimals and Percentages

These Fractions, Decimals and Percentages Are All the Same

The following fractions, decimals and percentages all mean the same thing.

They're really common, so it's a good idea to learn them.

$\frac{1}{2}$ is the same as 0.5, which is the same as 50%.

$\frac{1}{4}$ is the same as 0.25, which is the same as 25%.

$\frac{3}{4}$ is the same as 0.75, which is the same as 75%.

$\frac{1}{5}$ is the same as 0.2, which is the same as 20%.

$\frac{1}{10}$ is the same as 0.1, which is the same as 10%.

$\frac{1}{1}$ is the same as 1, which is the same as 100%.

You Can Change Fractions into Decimals

1) To convert a fraction into a decimal you should:

Divide the top number in the fraction by the bottom number.

EXAMPLE 1:

What is $\frac{9}{16}$ as a decimal?

Divide 9 by 16: $9 \div 16 = \mathbf{0.5625}$

EXAMPLE 2:

What is $\frac{3}{10}$ as a decimal?

Divide 3 by 10: $3 \div 10 = \mathbf{0.3}$

2) Use this method for trickier fractions when you don't have a calculator:

1) Multiply the top number by 10.

2) Divide by the bottom number.

3) Divide the result by 10.

EXAMPLE 3:

What is $\frac{4}{5}$ as a decimal?

1) Multiply the top number by 10: $4 \times 10 = 40$
2) Divide by the bottom number: $40 \div 5 = 8$
3) Divide by 10: $8 \div 10 = \mathbf{0.8}$

You Can Change Decimals into Fractions

To change a decimal into a fraction, you should:

1) Change the decimal to a whole number by multiplying by 10, 100 or 1000 etc.

2) Write this number as the top number of the fraction. Write the number you multiplied by as the bottom.

3) Simplify using equivalent fractions.

EXAMPLE:

What is 0.64 written as a fraction?
Give your answer in its simplest form.

1) Multiply the decimal by 100: $0.64 \times 100 = 64$

2) Write the result as the top number. The bottom number is 100: $0.64 = \dfrac{64}{100}$

3) Use equivalent fractions to write the fraction in its simplest form:

$$\dfrac{64}{100} \overset{\div 4}{\underset{\div 4}{=}} \dfrac{16}{25}$$

You can divide the top and bottom by 4.

Practice Questions

1) What is:

a) $\dfrac{3}{4}$ as a decimal?

b) 0.55 as a fraction?

2) What is:

a) 0.625 as a fraction?

b) $\dfrac{27}{45}$ as a decimal?

3) Mike runs $\dfrac{15}{16}$ of a mile. What is this as a decimal?

You Can Convert Between Fractions and Percentages

To change a fraction into a percentage, first change the fraction into a decimal (see page 35). Then multiply by 100 and add a % sign.

EXAMPLE 1:

What is $\frac{2}{5}$ as a percentage?

1) Change the fraction into a decimal: $2 \div 5 = 0.4$

2) Multiply by 100: $0.4 \times 100 = 40$

3) Add a % sign = **40%**

EXAMPLE 2:

28 out of 448 people surveyed are in favour of the closure of a local swimming pool. What percentage is this?

1) Write the number in favour as a fraction of the people surveyed: $\frac{28}{448}$

2) Change the fraction into a decimal: $\frac{28}{448} = 28 \div 448 = 0.0625$

3) Multiply by 100: $0.0625 \times 100 = 6.25$

4) Add a % sign = **6.25%**

To change a percentage into a fraction, write the percentage as the top number and 100 as the bottom number. Simplify using equivalent fractions.

EXAMPLE 3:

What is 85% as a fraction?

1) Write the percentage on the top and 100 on the bottom.

$$85\% = \frac{85}{100}$$

2) Use equivalent fractions to write the fraction in its simplest form.

$$\frac{85}{100} \overset{\div 5}{\underset{\div 5}{=}} \frac{17}{20}$$

EXAMPLE 4:

36% of a bar of white chocolate is cocoa. What is this as a fraction?

1) Write the percentage on the top and 100 on the bottom.

$$36\% = \frac{36}{100}$$

2) Use equivalent fractions to write the fraction in its simplest form.

$$\frac{36}{100} \overset{\div 4}{\underset{\div 4}{=}} \frac{9}{25}$$

Practice Questions

1) What is $\frac{108}{360}$ as a percentage?

....

2) What is 92% as a fraction?

....

3) What is $\frac{236}{400}$ as:

   a) a decimal? ....

   b) a percentage? ....

4) A bank carries out a survey into customer satisfaction. It finds that 250 out of 2000 customers are unhappy with the bank's service. What percentage is this?

....

....

Comparing Fractions, Percentages and Decimals

You need to be able to compare fractions, percentages and decimals.

EXAMPLE 1:

Which is greater, 0.44 or $\frac{7}{20}$?

1) You need to work out what $\frac{7}{20}$ is as a decimal.

2) To convert $\frac{7}{20}$ to a decimal, divide 7 by 20: $7 \div 20 = 0.35$

3) 0.35 is smaller than 0.44, so **0.44** is greater.

EXAMPLE 2:

Put these in order from smallest to largest: 0.56, 49%, $\frac{5}{8}$

Answer: Convert them all into the same form.

$49\% = 49 \div 100 = 0.49$ $\frac{5}{8} = 5 \div 8 = 0.625$

So the order is **49%, 0.56, $\frac{5}{8}$**.

You should write your answers in their original format.

EXAMPLE 3:

Niamh is looking at the offers on ready meals in the supermarket:

Sausage Supreme	Spaghetti Bolognese
Normally £2.55	Normally £2.00
Now **Two-Thirds of the Price!**	Special offer: **18% off**

Which meal works out cheaper for Niamh to buy?

First work out how much the Sausage Supreme will cost: $\frac{2}{3} \times 2.55 = £1.70$

Then work out how much the Spaghetti Bolognese will cost:

18% off = 100% − 18% = 82% = 0.82 $0.82 \times 2.00 = £1.64$

So the **Spaghetti Bolognese** with 18% off is cheaper.

Practice Questions

1) Circle the greater value in each pair.

   a) 0.04 or $\frac{6}{15}$?   b) 0.04 or 40%?   c) 0.3 or 3%?   d) $\frac{5}{8}$ or 67%?

....

....

....

2) LEX Comms normally offer broadband at £26.00 per month and line rental at £19 per month.   At the moment, they're offering the two deals below:

   Deal 1: 28% off broadband   Deal 2: two-fifths off line rental

   Which deal would save you the most money each month? Explain your answer.

....

....

....

....

Ratios

Ratios Compare One Part to Another Part

Ratios are a way of showing how many things of one type there are compared to another.

EXAMPLE:

A bathroom is decorated with 150 blue tiles and 180 white tiles.
What is the ratio of blue tiles to white tiles?

1) Write the relationship as a ratio: blue : white
                                        150 : 180

2) Simplify the ratio by dividing both sides
   by the same number each time.

$$150 : 180$$
$$\div 10 \searrow \qquad \swarrow \div 10$$
$$15 : 18$$
$$\div 3 \searrow \qquad \swarrow \div 3$$
$$5 : 6$$

So the ratio of blue tiles to white tiles is **5 : 6**.

Questions Involving Ratios

To answer a question involving ratios, you usually need to start by working out
the value of one part. For example, the cost of one thing or the mass of one part.
You can then use this to answer the question.

EXAMPLE 1:

Orange cordial is diluted by adding 4 parts water to every 1 part of
cordial (4 : 1). How much water should be added to 25 ml of cordial?

1) The amount of cordial used is 1 part, so 1 part = 25 ml.

2) You need 4 parts water to each part of cordial so:
   amount of water needed = 25 ml × 4 = **100 ml**

EXAMPLE 2:

Some jam is made from 1 part sugar to 3 parts fruit (1 : 3).
500 g of jam is made. How much sugar is used?

1) First work out how many parts there are in total.
   To do this, add up the numbers in the ratio: 1 + 3 = 4 parts

2) The jam contains 1 part sugar. To work out how many grams are in 1 part,
   divide the total amount of jam by the number of parts: 500 ÷ 4 = **125 g**

EXAMPLE 3:

£9000 is split between 3 people in the ratio 2:3:1.
How much money does each person get?

1) First work out how many parts are in the ratio.
   To do this, add up the numbers in the ratio.

$$2 + 3 + 1 = 6$$

2) To find out how much one part is worth, divide 9000 by 6: $9000 \div 6 = 1500$

3) The first person in the ratio gets two parts. To work out how much money
   they get, multiply the value of one part by 2:

$$1500 \times 2 = \textbf{£3000}$$

4) The second person in the ratio gets three parts. To work out how much
   money they get, multiply the value of one part by 3:

$$1500 \times 3 = \textbf{£4500}$$

5) The third person gets 1 part, so they get **£1500**.

To check your answer, make sure all the parts add up to £9000:
3000 + 4500 + 1500 = £9000.

Working Out Total Amounts

1) You can use ratios to work out total amounts.

2) You need to know the value of one part. (You may have to
   work this out or it might be given to you in the question.)

3) Then work out the total number of parts.

4) You can then multiply the total number of parts by the
   value of one part to find the total amount.

EXAMPLE:

A jelly is made from one part gelatin and four parts water.
320 g of water is used. How much jelly is made in total?

1) Find the value of one part by dividing the total amount of
   water by the number of parts of water: $320 \div 4 = 80$ g

2) Find the total number of parts by adding up
   the numbers in the ratio: $1 + 4 = 5$

3) Times the total number of parts by the amount
   given for one part: $5 \times 80 = \textbf{400 g}$

Practice Questions

1) Ollie is making salad dressing. He mixes **1** part vinegar to **2** parts oil.
   Ollie makes **450 ml** of salad dressing. How much oil does he use?

   ....

   ....

2) A union votes on whether to go on strike. **120** people vote.
   The ratio of yes : no votes is **2 : 3**.

   a) How many people vote yes?

   ....

   ....

   b) How many people vote no?

   ....

   ....

3) Hamish is making buttercream icing. He mixes **5** parts icing sugar to **3** parts butter.
   He uses **185 g** of icing sugar. How much butter will he need?

   ....

   ....

4) Dawn is mixing wallpaper paste. She mixes **3** parts glue to **7** parts water.
   Dawn uses **2.8** litres of water. How much wallpaper paste will she make in total?

   ....

   ....

5) George is a sheep farmer. His herd contains Herdwick sheep, Hebridean sheep
   and Soay sheep in the ratio **9 : 5 : 2**. George has **180** Herdwick sheep.
   How many sheep does he have in total?

   ....

   ....

   ....

Direct Proportion

You Can Use Proportion to Scale Up and Down

1) If an amount is directly proportional to another, it means that as one amount increases, the other amount increases at the same rate.

2) This means that if one amount is doubled, then the other amount is doubled as well.

3) You can use proportions to scale things up and down.

4) First, work out the value or amount needed for one thing by dividing.

5) Then multiply to get the value for the number of things you're looking for.

EXAMPLE 1:

Lucy is making cakes. She finds this recipe. ➡️

Lucy wants to make 20 cakes.
How much margarine does she need?

Recipe for 12 cakes:
150 g flour
75 g sugar
75 g margarine
3 eggs

1) Start by dividing to work out how much margarine is needed for 1 cake.

   The recipe is for 12 cakes, so you need to divide the weight of the margarine by 12:
   $$75 \div 12 = 6.25 \text{ g}$$

2) Multiply the weight of margarine needed for 1 cake by 20 to find out how much margarine she needs for 20 cakes:
   $$6.25 \times 20 = \textbf{125 g}$$

EXAMPLE 2:

A breakfast cereal contains 0.4 g of calcium per 100 g.
How much calcium does a 35 g serving of the breakfast cereal contain?

1) Start by dividing to work out how much calcium is in 1 g of the cereal.

   There is 0.4 g of calcium in 100 g of cereal, so you need to divide 0.4 by 100:
   $$0.4 \div 100 = 0.004 \text{ g}$$

2) Multiply your answer by 35 to find out how much calcium is in 35 g of cereal:
   $$0.004 \times 35 = \textbf{0.14 g}$$

EXAMPLE 3:

Rhys is a baker. He bakes 500 biscuits in a 4-hour shift.
How many biscuits can Rhys bake in a 6-hour shift?

1) Start by dividing to work out how many biscuits Rhys bakes in 1 hour.

He bakes 500 biscuits in 4 hours, so you need to divide 500 by 4:
$$500 \div 4 = 125 \text{ biscuits}$$

2) Multiply to find how many biscuits Rhys can bake in 6 hours:
$$125 \times 6 = \textbf{750 biscuits}$$

Practice Questions

1) Freya is making soup. She needs 500 g of carrots to make 1 litre of soup.
How many grams of carrots does she need to make 1.5 litres of soup?

....

....

2) 1000 ml of lemonade contains 250 ml of lemon juice.
How much lemon juice does 600 ml of lemonade contain?

....

....

3) Leon runs 10 km in 49 minutes. Assuming he runs at the same speed,
how long should it take him to run 18 km?

....

....

4) Srishti collects the following information about train fares and journey distance.

Price of Ticket (£)	4	8	12	18
Train Journey Distance (km)	16	32	64	90

She says that the distance of a train journey and the price of the ticket
are directly proportional. Is she correct? Explain your answer.

....

....

....

Inverse Proportion

Inverse Proportion is the Opposite of Direct Proportion

1) If an amount is inversely proportional to another, it means that as one amount increases the other decreases at the same rate.

2) This means that if one amount is doubled, then the other amount is halved.

3) To work out how a value changes, first you need to multiply to find the value for one thing.

4) Then, you need to divide to find the value you're looking for.

EXAMPLE 1:

5 people take 12 minutes to change all of the light bulbs in a restaurant. How long would it take 3 people to change all of the light bulbs?

1) Start by working out how long it would take one person to change the light bulbs.

   5 people take 12 minutes, so you need to multiply 5 by 12:
   $$5 \times 12 = 60 \text{ minutes}$$

2) Divide to find the time for 3 people:
   $$60 \div 3 = \textbf{20 minutes}$$

EXAMPLE 2:

2 people will each have to paint at a rate of 45 m² per hour to finish painting a house exactly on schedule.

At what rate would 6 people each have to paint to finish exactly on schedule?

1) Start by working out the rate for one person. There are 2 people so multiply 45 by 2:

   $$2 \times 45 = 90 \text{ m}^2 \text{ per hour}$$

2) Then divide to find the rate for 6 people:

   $$90 \div 6 = \textbf{15 m}^2 \textbf{ per hour}$$

How many people would be needed so they could paint at a rate of 9 m² per hour?

3) Divide the rate for one person by the new rate (9) to find the number of people needed:

   $$90 \div 9 = \textbf{10 people}$$

EXAMPLE 3:

Peter is making solid balls out of different metals. The balls all have the same weight. The volume of each ball is inversely proportional to its density.

He makes one ball out of aluminium. Its volume is 108 cm³ and its density is 2.6 g/cm³. He uses steel to make a new ball with a density of 7.8 g/cm³.

What is the volume of the steel ball?

See page 69 to find out more about density.

1) First, look at how the densities are related.

$$7.8 \div 2.6 = 3$$

So the density of the steel ball is three times greater than the density of the aluminium ball.

2) Since volume is inversely proportional to density, the volume of the steel ball must be a third of the volume of the aluminium ball.

$$108 \div 3 = \textbf{36 cm}^3$$

Practice Questions

1) A call centre has 5 people answering calls. To answer all of the calls, each person needs to work at a rate of 14 calls per hour. At what rate would 7 people need to work to answer the same number of calls?

....

....

....

2) The time is takes an object to move a certain distance is inversely proportional to the speed that it's travelling at.

It takes a car 12 minutes to travel the length of a road if it goes at an average speed of 60 mph.

a) How long would it take the car to travel down the same road if it travelled at an average speed of 30 mph?

....

....

b) If it takes the car 36 minutes to drive down the road, what will its average speed be?

....

....

Formulas in Words

A Formula is a Type of Rule

1) A formula is a rule for working out an amount.

2) Formulas can be written in words. Sometimes, it can be tricky to spot the formula.

EXAMPLE:

Sia packs 40 boxes an hour. How many boxes can she pack in 6.5 hours?

You're told that: "Sia packs 40 boxes an hour." This is a formula. You can use it to work out how many boxes Sia can pack in a given number of hours.

1) The calculation you need to do here is:

Number of boxes = 40 × number of hours

2) You've been asked how many boxes Sia can pack in 6.5 hours, so put '6.5' into the calculation in place of 'number of hours':

Number of boxes = 40 × 6.5 = **260**

You can use the same formula to work out how many boxes Sia can pack in any number of hours.

Formulas Can Have More Than One Step

Some formulas have two steps in them. You need to be able to use two-step formulas.

EXAMPLE:

Owen has moved into a new house. The telephone company will charge him £110 to connect his phone line, then line rental at £11.50 per month. How much will Owen's phone line have cost him after 12 months?

The formula here is "£11.50 per month, plus £110".

1) Work out the calculation you need to do:

Step 1: 11.50 × number of months

Step 2: + 110

There's more on calculations involving brackets on page 10.

Cost of phone line = (11.50 × number of months) + 110

2) Then just put the right numbers in.
In this case it's '12' in place of 'number of months':

(11.50 × 12) + 110 = **£248**

Practice Questions

1) Dan is getting some furniture delivered. Delivery costs £5 per item.
   How much will it cost Dan to get 7 items of furniture delivered?

   ....

   ....

2) Chrissie needs a wallpaper stripper. It costs £12.50 a day to hire, plus a deposit of £40.
   How much will it cost Chrissie to hire the wallpaper stripper for 3 days (and the deposit)?

   ....

   ....

3) Angela is leaving her car parked at the airport whilst she goes away on business for 5 days.
   It costs £24.50 per day to park there and there is a one-off charge of £15.
   How much will it cost Angela to park for 5 days?

   ....

   ....

4) Shabnam is a babysitter. She charges £5.00 an hour before midnight and £6.50 an hour
   after midnight. How much will Shabnam earn babysitting from 8 pm to 2 am?

   ....

   ....

   ....

5) Josie takes a taxi when she travels between her house and town.
   The journey from her house to town (or the other way around) usually costs £8.
   If she travels after 11 pm the journey costs £12.

   In one week Josie travelled to town and back four times. All the trips to town were before
   11 pm. One journey back was after 11 pm. How much did she spend on taxis?

   ....

   ....

   ....

Formulas Using Letters

Formulas Can be Shown Using Letters

1) You might be given a formula made up of letters. Each letter represents something.

2) Sometimes formulas don't use '×' or '÷'.

3) Instead, $a \times b$ is written as ab and $a \div b$ is written as $\frac{a}{b}$.

> To work out the surface area of a hemisphere, you square the radius and multiply by three lots of pi.
>
> As words, this can be written as:
>
> surface area $= 3 \times \pi \times$ radius^2
>
> The formula can be shortened by using letters instead of full words.
>
> $$S = 3 \times \pi \times r^2 = 3\pi r^2$$
>
> S represents surface area r represents radius
>
> For more on surface area see page 88 and for pi see page 73.

Putting Numbers into Formulas

If you have a formula written in letters, you can substitute numbers in by replacing the letters with numbers.

EXAMPLE 1:

A bowl of jelly has a radius of 9 cm. When the jelly is tipped out of the bowl it is a perfect hemisphere. What is its surface area? Give your answer to the nearest whole number.

 $S = 3\pi r^2$

1) Start by working out which letter you know the value of.

   You're told that the radius is 9 cm, so $r = 9$ cm.

2) Then substitute this number into the formula:

   $S = 3\pi r^2$

   $S = 3 \times \pi \times r^2$

   $S = 3 \times \pi \times 9^2$

   The units are cm squared (cm^2) as you're multiplying a radius in cm by a radius in cm.

   $S = 763.4... = 763$ to the nearest whole number

So the surface area of the jelly is **763 cm^2** to the nearest whole number.

EXAMPLE 2:

Temperature can be measured in degrees Celsius (°C) or degrees Fahrenheit (°F).
The formula $F = \frac{9}{5}C + 32$ can be used to swap between the two.
C represents Celsius and F represents Fahrenheit.
If the temperature is 15 °C, what is the temperature in Fahrenheit?

1) Write out the formula in full: $\qquad\qquad\qquad$ $F = \frac{9}{5} \times C + 32$

2) Put numbers in place of any letters you know.
   Here you're told that C = 15: $\qquad\qquad$ $F = \frac{9}{5} \times 15 + 32$

3) Work it out in stages.
   Write down values for each bit as you go along: $\quad$ $F = 27 + 32$
   $\qquad\qquad\qquad\qquad\qquad\qquad\qquad\qquad\qquad\qquad$ $F = 59$

So the temperature is **59 °F**.

EXAMPLE 3:

GlobalPhone mobile phone contracts are worked out using the formula below:

$$\text{cost} = m(5n + 1000)$$

cost = price in pence, m = number of months,
n = number of text messages per month.

Rosie buys a 12-month contract, with 500 texts per month.
She also buys a new phone. In total, the contract and the phone cost £780.
What is the price of the new phone?

1) Start by writing the formula out in full.
   The formula can be written as: $\quad$ cost $= m \times (5 \times n + 1000)$

2) Work out what values the letters have:
   $\qquad\qquad$ Rosie buys a 12-month contract, so $m = 12$.
   $\qquad\qquad\qquad\qquad$ She gets 500 texts, so $n = 500$.

3) Substitute these numbers into the formula:
   $\qquad\qquad$ cost $= 12 \times (5 \times 500 + 1000)$
   $\qquad\qquad$ cost $= 12 \times (3500)$
   $\qquad\qquad$ cost $= 42\,000\text{p} = £420$

4) Find the price of the phone by subtracting the
   cost of the contract from the total cost:
   $\qquad\qquad$ $780 - 420 = £360$

So the price of the new phone is **£360**.

Practice Questions

1) What is the value of $6ab^2$ if a is 4 and b is 7?

   ....

2) What is the value of $\frac{4a}{b}$ when $a = 9$ and $b = 12$?

   ....

3) A teacher is organising a school trip. The formula $s = c \div 20$ is used to work out the number of staff needed on the trip. s represents the number of staff needed and c represents the number of children on the trip. If 80 children are on the trip, how many staff are needed?

   ....

   ....

4) The area of a trapezium is given by the length of the base plus the length of the top, multiplied by half of the height.

   a) Write this as a formula where A = area, a = top length, b = base length and h = height

   ....

   b) What is the area of a trapezium with a base length of 13 cm, a top length of 5 cm and a height of 6 cm?

   ....

5) Aki uses the formula below to work out how many fence posts he needs.

   $$n = \frac{p}{8} + 5$$
   n = number of fence posts p = number of fence panels

   Aki's front garden fence will have 112 panels. How many fence posts does he need?

   ....

   ....

6) The formula on the right is used to find the cost in pounds of hiring one bike (including a deposit). Jill wants to hire 4 bikes for 5 hours. How much will this cost?

Cost = $10n + 5$
n = number of hours

   ....

   ....

Section Two — Measures, Shape and Space

Money

Working with Money

1) If you get a question on money, the units will probably be pounds (£) or pence (p).

2) You need to be able to switch between using pounds and using pence.

> To go from pounds to pence, multiply by 100.
>
> To go from pence to pounds, divide by 100.

Remember that
£1 = 100p.

3) You may get a question that uses pounds and pence.

4) If you do, you'll need to change the units so that they're all in pounds or all in pence.

EXAMPLE:

Callum buys fish and chips for £5.25, a carton of gravy for 60p and a cup of tea for 75p. How much does he need to pay in total?

1) Change the carton of gravy and cup of tea from pence to pounds.

$$60p ÷ 100 = £0.60$$
$$75p ÷ 100 = £0.75$$

2) All the prices are in the same units now (£), so just add them up.

$$£5.25 + £0.60 + £0.75 = £6.60$$

5) If the question tells you what units to give your answer in then make sure you use those. If it doesn't, then it's a good idea to use pounds if the answer is more than 99p.

Practice Questions

1) a) What is £16.42 in pence?   b) What is 210p in pounds?

   ....

2) Nasir buys a badminton set for £27.99, a sweatband for £1.20 and a bottle of water for 70p. He pays with a £10 and a £20 note. How much change does he get?

   ....

   ....

   ....

Discounts or Increases Can Be Given as Percentages

Money questions often involve percentages.

See page 28 for more on working out percentages with or without your calculator.

For example, you might have to work out how much you save when you get a discount. You could also be asked the new price of something when it is increased or reduced by a certain percentage.

EXAMPLE 1:

A shop is having a clearance sale. All items are 42% off.
A pair of designer shoes normally costs £325.

1) How much do you save when buying the shoes during the sale?

   Write it down: you save 42% of 325.

   Change the percentage into a decimal: 42% = 42 ÷ 100 = 0.42

   Change the 'of' to '×' and work it out: 42% of £325 = 0.42 × 325
   = **£136.50**

   Don't forget to use the correct money format. (The answer is £136.50, not £136.5).

2) How much do the shoes cost during the sale?

   Subtract the savings from the normal cost: £325 − £136.50 = **£188.50**

EXAMPLE 2:

A camp site charges £54 per day to park a caravan. The price is increased by 18% during the peak season. This calendar shows the dates of the peak season.

How much does it cost to park a caravan for 7 days beginning on Monday the 21st?

Mo	Tu	We	Th	Fr	Sa	Su
	1	2	3	4	5	6
7	8	9	10	11	12	13
14	15	16	17	18	19	20
21	22	23	24	25	26	27
28	29	30	31			

☐ = peak season

1) Find the cost per day of parking in peak season.

   Find 18% of £54: 0.18 × 54 = £9.72

   Add this to £54: 54 + 9.72 = £63.72

   You could do this in one step by calculating 118% of the normal cost:
   118% of £54 = 1.18 × 54 = £63.72

   So in peak season, it costs £63.72 per day to park a caravan.

2) Now work out the total cost.

   Out of the 7 days there are...

   - 4 days not in peak season: 4 × £54 = £216
   - 3 days in peak season: 3 × £63.72 = £191.16

   So in total it costs 216 + 191.16 = **£407.16** to park the caravan.

Discounts Involving Fractions

You might need to calculate a fraction of a price to work out a discount.

See page 20 for how to find fractions of amounts.

EXAMPLE:

A cruise usually costs £1698. A special offer gives $\frac{5}{12}$ off the price. What is the special offer price of the cruise?

1) First you need to work out $\frac{5}{12}$ of £1698:

$5 \div 12 \times 1698 = £707.50$

So the special offer gives £707.50 off the cruise.

2) Then you need to take this number away from £1698:

$1698 - 707.50 = £990.50$

So the special offer price is **£990.50**.

Practice Questions

1) A pair of boots normally cost £152 but are on sale with $\frac{3}{8}$ off of their normal price. How much do they now cost?

....

2) A fast food chain buys burgers from a food supplier. They're given a discount depending on how many they buy. The table on the right is used to calculate this discount.

The full price of a single burger is 27p.

What is the difference in the price to pay between ordering 1100 burgers as a single order or as two orders of 550? Give your answer in pounds (£).

Quantity	Discount
Fewer than 500	0
500 to 1000	$\frac{2}{15}$
More than 1000	$\frac{3}{15}$

....

....

....

3) Frankie is deciding which of three new cars to buy. The table on the right shows the discounts she would get on each car.

Which car is the cheapest after the discounts?

Car	A	B	C
Price	£6872	£6611	£6480
Discount	$\frac{3}{8}$	$\frac{3}{11}$	$\frac{2}{9}$

....

....

Profit as a Percentage

1) Profit is the difference between your costs and the money that you make.

2) The profit from selling an item is often given as a percentage of the costs involved.

EXAMPLE:

Chloe makes and sells stuffed animal toys. The cost of making her last toy was £40. The toy sold on an auction website for £52.
How much profit did she make as a percentage of her costs?

1) Work out the profit: Profit = selling price – costs
                                              = £52 – £40 = £12

2) Work out the percentage profit: % profit = 12 ÷ 40 × 100 = **30%**

Practice Questions

1) Sabrina buys a raincoat. It normally costs £59, but she buys it at a 12% discount. How much does Sabrina save using the discount?

....

2) A parking fine costs £70. If it is not paid within two weeks, it increases by 62%. How much will the parking fine cost after two weeks?

....

3) Tahel works in a supermarket. She is marking down items for a sale. A TV costing £248 is given a 26% discount. Tahel prices the TV at £194. Has Tahel changed the price correctly?

....

....

4) Dafydd sells DVDs on a market stall. He buys the DVDs for £2.60 and sells them for £3.90. What is Dafydd's percentage profit on each DVD?

....

5) Luke makes fruit cakes for a stall at a village fete. It costs Luke £1.80 for the ingredients for each cake. If he wants to make exactly 35% profit on each cake, how much money should Luke charge for each cake?

....

....

Calculate the Price Per Item to Work Out Value For Money

1) If you're buying a pack of something, you can work out how much you're paying for each item.

Price per item = total price ÷ number of items

2) You can then compare the price per item for that pack with other packs.

EXAMPLE:

A shop sells crisps in multipacks of 6 or 11.
The 6-pack costs £1.68. The 11-pack costs £2.97.
Which pack is better value for money?

6-pack: Price per bag = £1.68 ÷ 6 = £0.28

11-pack: Price per bag = £2.97 ÷ 11 = £0.27

Price per bag = total price ÷ number of bags

The **11-pack** costs less per bag, so it's better value than the 6-pack.

Work Out the Value of Offers to Find the Best Deal

To find the best deal, you need to work out how much you'd pay with each offer or which offer takes the most off the total price.

EXAMPLE:

Malcolm wants to hire a van for five days from either Hire-a-van or Speedy Hire.

Hire-a-van — £165 per day with 15% off the total price
for bookings of more than one day

Speedy Hire — £163 per day

Malcolm has a voucher for £99 off that he could use with Speedy Hire.

Which company should Malcolm use to get the best deal?

Hire-a-van ⟶ Cost of five days' hire = £165 × 5 = £825
Discount for booking five days = 0.15 × £825 = £123.75
Total cost = £825 − £123.75 = £701.25

You could also do 15% off = 100% − 15% = 85% of the price, and 0.85 × £825 = £701.25.

Speedy Hire ⟶ Cost of five days' hire = £163 × 5 = £815
Voucher discount = £99
Total cost = £815 − £99 = £716

So Malcolm will get the best deal by hiring a van from **Hire-a-van**.

Practice Questions

1) A pet shop sells a 6-can multipack of dog food for £3.18 and a 15-can multipack for £7.80. Which multipack is the best value for money?

....

....

2) Gillian is buying a new carpet that she would like fitting in her lounge. The carpet showroom offers Gillian free fitting or 20% off her total bill.

The cost of the carpet is £396.50. The cost of fitting is £120.00. Which offer will save Gillian the most money?

....

....

3) Bottles of lemonade are sold individually and in packs of four. A pack of four bottles normally costs £4.80, but today there is a 50% discount. Individual bottles are on a 'buy one, get one half price' offer. They normally cost £0.90 each.

Josh wants to buy 12 bottles of lemonade. What is the cheapest way that he can do so?

....

....

....

Interest

1) Interest is money that's added on to the value of something. It's given as a percentage.

2) For example, money saved in a bank account earns interest. Items that you buy on payment plans also cost more over time because interest is charged on them.

3) When working with interest, calculate the percentage of the amount and then add it to the original amount.

> **EXAMPLE:**
>
> Lara takes out a loan of £10 950 to pay for a car. She pays back the loan in one year, plus 12% interest. How much money did Lara pay back in total?
>
> Find 12% of £10 950: $0.12 \times 10\,950 = £1314$
>
> Add this on to £10 950: $10\,950 + 1314 = $ **£12 264**
>
> You could also do this in one step: $1.12 \times 10\,950 = £12\,264$.

Compound Interest is 'Interest on Interest'

1) Interest can be added over and over again. For example:

- Money saved in a bank account might earn interest once every year.

- Money owed on a credit card might be charged interest every month.

2) Compound interest is interest that is calculated as a percentage of the current value of something, including any interest that has already been applied.

EXAMPLE:

Yusuf puts £40 into a savings account that earns 5% compound interest per annum. How much money will he have after three years? *'per annum' means once a year.*

You need to add on interest once a year for three years.
Each year the interest is calculated as 5% of the current savings.

1) Find 5% of £40: $0.05 \times 40 = £2$
Add this to £40: $40 + 2 = £42$

Compound interest increases every year, but it is always 5% of the current amount.

So Yusuf will have £42 after one year.

2) Find 5% of £42: $0.05 \times 42 = £2.10$
Add this to £42: $42 + 2.10 = £44.10$

Round to the nearest 1p if the interest has more than 2 decimal places.

So Yusuf will have £44.10 after two years.

3) Find 5% of £44.10: $0.05 \times 44.10 = £2.205 = £2.21$ to 2 d.p.
Add this to £44.10: $44.10 + 2.21 = £46.31$

So Yusuf will have **£46.31** after three years.

Practice Questions

1) Frank puts £1825 into a pension fund that earns 6% compound interest per annum. How much is the pension fund worth after 2 years?

....

....

2) Zola takes out a loan of £750. She is charged 10% compound interest every month. If Zola doesn't pay back any money, how much will she owe after three months?

....

....

....

Budgeting

1) A budget is a record of the amount of money you earn and the amount you need to spend.

2) Budgets are used to control costs and incomes. For example, a business might use a budget so that it doesn't spend more than it earns.

3) A budget can be a list or table of things with an amount of money set aside for each.

Hafeez plans to open a market stall. The table on the right shows his budget for weekly costs. What percentage of his costs are in rent?

Item	Amount (£)
Stock	705
Rent	217
Wages	163

Answer: 1) Find the total costs for a week:

705 + 217 + 163 = £1085

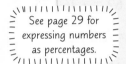
See page 29 for expressing numbers as percentages.

2) Find the cost of rent as a percentage of the total:

217 ÷ 1085 × 100 = **20%**

Hafeez calculates that he'll need to earn £155 on each day of the week to break even. Is Hafeez correct?

To "break even" means to have a total income equal to total cost.

Answer: There are 7 days in a week, so his weekly income would be: 7 × £155 = £1085

His weekly costs are £1085. So **yes** — Hafeez is correct.

Practice Questions

1) Deborah is planning a holiday. Her budget is shown in the table below.

Item	Travel	Hotel	Shopping	Food & Drink
Amount (£)	416	256	384	144

a) What is Deborah's total budget?

....

b) What percentage of Deborah's total budget is set aside for shopping?

....

2) A social club is organising its annual banquet.
The club has 120 members and at least a quarter are expected to attend.
A catering company is hired to supply the food. They charge £24.50 per person.
What is the minimum amount of money that the club should budget for the food?

....

....

Rates of Pay

1) A rate of pay is how much something costs per unit of time. For example, a phone call might cost 5p per minute. If a call lasted for 8 minutes, it would cost 8 × 5p = 40p.

2) Rates of pay can also be used to show how much money someone earns. For example, a person might earn £7.83 per hour.

EXAMPLE:

Jill is supposed to work 37.5 hours a week and earns £9.62 per hour.
During a busy week, she is asked to work a total of 40 hours.
For the extra hours, her pay is increased to £12.10 per hour.

How much will Jill earn this week?

1) Find how much she earns for her normal hours: 37.5 × £9.62 = £360.75

2) Calculate how many extra hours she will work: 40 − 37.5 = 2.5 hours

3) Find how much she earns for the extra hours: 2.5 × £12.10 = £30.25

In total, Jill will earn 360.75 + 30.25 = **£391**.

Practice Questions

1) A premium-rate telephone line charges £3.60 per minute, plus a fixed cost of £1.20. How much would a call lasting half an hour cost?

....

....

2) Ahmed has a job that pays £410 a week. How much will Ahmed earn in:

a)   12 weeks?   b)   52 weeks?

....

Ahmed is offered a new job that pays £22 000 per year.

c)   If Ahmed takes this new job, will he earn more in a year than with his current job?

....

3) Leanna hires a cleaning company to clean her house. They charge £15 per hour. She was charged £48.75. How long did it take to clean Leanna's house?

....

....

Taxes

1) A tax is an amount of money paid to the government.

2) For example, council tax is paid to the local council.
   The amount paid depends on the value of the property that you live in.

EXAMPLE 1:

The table on the right is used to calculate council tax.
Marv's house is valued in Band C. He lives alone,
so the council tax is reduced by 25%.
How much does Marv pay in council tax?

Band	Council tax
A	£1035
B	£1208
C	£1380
D	£1553

1) The council tax for Band C is £1380.

2) Find 25% of £1380:   0.25 × 1380 = £345

3) Subtract this from £1380:   1380 − 345 = **£1035**

You could do: 25% reduction
= 100% − 25%
= 75% of cost
Then 0.75 × £1380 = £1035.

3) Another example is income tax. It is calculated as a percentage of a
   person's income — but only the income that's above a certain amount.

EXAMPLE 2:

Jameela must pay 20% income tax on all earnings above £12 500 in a year.
If Jameela earns £35 000 in one year, how much income tax does she pay?

1) Work out how much Jameela earns above £12 500:

   35 000 − 12 500 = £22 500

2) Now calculate 20% of the remaining amount:

   20% of £22 500 = 0.2 × 22 500 = **£4500**

The amount of money
that isn't taxed is called
a 'personal allowance'.
Here, it's £12 500.

Practice Question

1) VAT (value-added tax) is paid on things that you buy.
   The table on the right shows the 2019 VAT rates.
   This is how much VAT is charged on certain items
   as a percentage of the item's cost.

VAT (%)	Items
20	Chocolate and crisps
5	Gas and electric
0	Fruit and vegetables

Before VAT is added, Simon pays 12p per unit of
electricity plus a fixed charge of £87 per year.
How much does Simon pay in VAT if he uses 3000 units of electricity in one year?

....

....

....

Units

Measures Have Units

1) Things that you measure have units. For example, metres (m) or grams (g).

2) They're really important. For example, you can't just say that a distance is 4 —
you need to know if it's 4 metres, 4 kilometres, 4 miles, etc.

3) There are two types of units: metric and imperial. You'll need to be able to use both.

Units of Length

1) Length is how long something is.

2) Metric units for length are millimetres (mm),
centimetres (cm), metres (m) and kilometres (km).

3) Imperial units for length are inches (in),
feet (ft), yards (yd) and miles.

4) Length can be measured in either metric or imperial units.
For example, a football pitch may have a length of 90 m,
which is about the same as 98 yd.

Metric length
1 cm = 10 mm
1 m = 100 cm
1 km = 1000 m

Imperial length
1 foot = 12 inches
1 yard = 3 feet
1 mile = 1760 yards

Units of Weight

1) Weight is how heavy something is.

2) Metric units for weight are grams (g) and kilograms (kg).

3) Imperial units for weight are ounces (oz),
pounds (lb) and stones (st).

4) Weight can be measured in either metric or imperial units.
For example, a 7 pound baby weighs about 3.2 kg.

Metric weight
1 kg = 1000 g

Imperial weight
1 lb = 16 oz
1 stone = 14 lb

Units of Capacity

1) Capacity is how much something will hold.

2) Metric units for capacity are millilitres (ml) and litres (L).

3) Imperial units for capacity are fluid ounces (fl. oz),
pints (pt) and gallons (gal).

4) Capacity can be measured in either metric or imperial units.
For example, a 1 litre bottle holds about 1.76 pints.

Metric capacity
1 L = 1000 ml

Imperial capacity
1 pint = 20 fl. oz
1 gallon = 8 pints

Use a Conversion Factor to Convert Between Units

1) To convert between units, you need to multiply or divide by some number. This number is called the conversion factor. For example:

> 1 km = 1000 m
> To convert between kilometres and metres, the conversion factor is 1000.

> 1 foot = 12 inches
> To convert between feet and inches, the conversion factor is 12.

2) Whatever units you're using, the method is always the same:

- Find the conversion factor.
- Decide whether to multiply or divide by it.
- Work out your answer.

3) For metric to metric conversions, the conversion factor will be 10, 100 or 1000. You'll need to remember these — you won't be told them in the test.

EXAMPLE:

What is 2.7 km in m?

Metric length
1 km = 1000 m

1) Write out the information that you know:
   1 km = 1000 m
   So the conversion factor is 1000.

2) A kilometre is longer than a metre. So to go from km to m, the number should get bigger. So you need to multiply.

3) Work out your answer: 2.7 × 1000 = **2700 m**

Practice Questions

1) How many metres are in 7.5 km?

....

2) What is 6400 g in kg?

....

3) What is 560 millilitres in litres?

....

Conversions With Imperial Units

1) If you have to convert imperial units, you'll be told the conversion factor.

2) You could get imperial to imperial conversions as well as conversions between imperial and metric units.

EXAMPLE 1:

1 gallon is equal to 8 pints.
How many pints are in 13 gallons?

1) You're told that 1 gallon = 8 pints.
   So the conversion factor is 8.

2) A gallon is more than a pint.
   So to go from gallons to pints,
   the number should get bigger.
   So you need to multiply:

   13 gallons is 13 × 8 = **104 pints**

EXAMPLE 2:

1 oz is equal to 28 g.
How many ounces are in 140 g?

1) You're told that 1 oz = 28 g.
   So the conversion factor is 28.

2) A gram is less than a ounce.
   So to go from g to oz, the
   number should get smaller.
   So you need to divide:

   140 g is 140 ÷ 28 = **5 oz**

Practice Questions

1) 1 kg is equal to 2.2 lb. How many lbs are in 15 kg?

....

2) 1 km = 0.62 miles. Sally has cycled 10 km. How far has she cycled in miles?

....

3) 1 litre = 35 fluid ounces (fl. oz). Joe needs 7 fl. oz of milk for a recipe.

   a) How many litres of milk does Joe need?

   ....

   ....

   b) How many millilitres of milk does Joe need?

   ....

4) 1 foot = 12 inches. 1 inch = 2.54 cm.
   Chris is 6 feet and 3 inches tall. How tall is he in cm?

....

....

Converting Between Other Units

1) You can convert between any units so long as you know the conversion factor.

2) For example, you can convert between currencies. If the exchange rate between pounds (£) and euros (€) is £1 = €1.13, then the conversion factor is 1.13.

3) Another example is converting between millilitres and cubic centimetres: 1 ml = 1 cm³, so the conversion factor is 1.

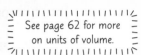
See page 62 for more on units of volume.

EXAMPLE:

Fiona visits Japan and returns home with 12 000 Japanese Yen (¥). How much is this in pounds? The exchange rate is £1 = ¥137.35.

1) The conversion factor is 137.35.

2) 1 Yen is worth less than £1. So to go from Yen to pounds, the number should get smaller. So you need to divide.

3) So ¥12 000 is equal to 12 000 ÷ 137.35 = 87.368... = **£87.37**.

Practice Questions

1) 1 British pound (£) is worth 1.2 US dollars ($).
   Deepika paid $450 for a flight. How much is this in pounds?

   ....

   ....

2) Arnie has an empty jug with a capacity of 500 cm³. He pours water into the jug, stopping when the water takes up three quarters of the jug's capacity.

   How many millilitres of water did Arnie pour into the jug? (1 cm³ = 1 ml)

   ....

   ....

3) Jess has a swimming pool in the shape of a cuboid.
   It is 4 m wide, 9 m long and has a depth of 2 m.

   How many gallons of water are needed to fill Jess's pool?
   (1 m³ = 1000 L and 1 gallon = 4.5 L)

   2 m

   9 m

   4 m

   ....

   ....

Conversion Graphs

A conversion graph is a line graph that can be used to convert between units.

EXAMPLES:

Here's a conversion graph for miles and kilometres.

1) How many miles is 8 km?

- Find 8 km on the horizontal axis and move up until you get to the line.
- Go directly across to the vertical axis.
- The value on the vertical axis is the answer — **5 miles**.

2) How many km is 3 miles?

- Find 3 km on the vertical axis and move across until you get to the line.
- Go straight down to the horizontal axis.
- The value on the horizontal axis is the answer, so it's roughly **5 km**.

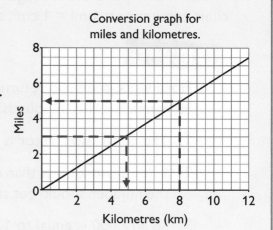

Conversion graph for miles and kilometres.

Practice Questions

1) Use the conversion graph in the example above to answer this question.

   a) What is 6.5 km in miles?

   b) Which length is greater: 7 miles or 11 km?

   ....

   ....

2) The graph on the right can be used to convert between pounds and kilograms.

   a) What is 2.2 lb in kg?

   ....

   b) What is 1.8 kg in lb?

   ....

   Conversion graph for pounds and kilograms.

   c) Greg is baking bread.
   He needs to weigh 2 kg of flour, but his scales only show pounds. How many pounds of flour should he weigh?

   ....

Speed and Density

Speed is 'distance per unit of time'

1) Speed is how fast something is moving. Some common units for speed are miles per hour (mph), metres per second (m/s) and kilometres per hour (km/h).

2) To calculate the speed of something, you need to divide the distance it has moved by the time it has taken to move that distance.

3) The formula for calculating speed is: speed = distance ÷ time

EXAMPLE 1:

Eric drives 180 miles in 3 hours. What is his speed in mph?

1) Write out the formula: speed = distance ÷ time

2) Put in the numbers: speed = 180 ÷ 3

3) Work out the answer: speed = **60 mph**

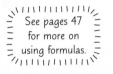
See pages 47 for more on using formulas.

4) You may need to convert one or more of the units first to get the correct unit of speed.

EXAMPLE 2:

Siobhan runs 9 km in 45 minutes. What is her speed in km/h?

1) You're asked for the speed in kilometres per hour (km/h) but the time is given in minutes. So first, convert from minutes to hours:
45 minutes $= \frac{3}{4}$ of an hour $= 0.75$ hours

2) Now work out the speed using the formula:
speed = distance ÷ time = 9 ÷ 0.75 = **12 km/h**

Practice Questions

1) Liam runs 200 m in 25 seconds. What is his speed in m/s?

....

2) Tina completes a 26-mile marathon in 150 minutes. What was her speed in mph?

....

....

Distance and Time Calculations

If you know the speed and either the distance or the time,
then you can work out the one that you don't know.

The formula for time is:
time = distance ÷ speed

The formula for distance is:
distance = speed × time

EXAMPLES:

A train is moving at a speed of 10 m/s.

1) How much time does it take for the train to travel 50 m?

   Use the formula: time = distance ÷ speed

   = 50 ÷ 10 = **5 seconds**

   The speed is given in m/s (metres per second) so the time is in seconds.

2) How far does the train travel in 7 seconds?

   Use the formula: distance = speed × time

   = 10 × 7 = **70 m**

   The speed is given in m/s (metres per second) so the distance is in metres.

Practice Question

1) Dimitri is planning a car journey.
   He has sketched the roads that he could take.

   Dimitri will obey the following speed limits:

   - The road from A to B has a 50 mph speed limit.
   - The road from B to C has a 20 mph speed limit.
   - The road from A to C has a 70 mph speed limit.

   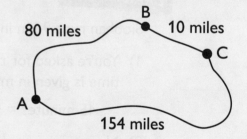
   80 miles   B   10 miles   C   A   154 miles

   a) What is the fastest route from A to C?

   ....

   ....

   ....

   b) Dimitri decides to take the route that passes through B.

   He sets off from point A. After $1\frac{1}{2}$ hours, his car breaks down.

   If he travels within the speed limit, what is the farthest he could have travelled?

   ....

   ....

Density

1) Density is a measurement that relates the weight of something to its volume. Some common units of density are g/cm³ (grams per cubic centimetre) and kg/m³ (kilograms per cubic metre).

2) For example, if you had equal sized cubes of cork and lead, then the cork would weigh less. It has a lower density.

3) The formulas linking density, weight and volume are:

   You might see weight referred to as 'mass', but just treat it the same.

> density = weight ÷ volume
>
> volume = weight ÷ density
>
> weight = density × volume

EXAMPLE 1:

Yoshi puts a measuring jug on top of some weighing scales and sets the scale to zero. He then pours water into the jug. What is the density (to 2 decimal places) of the water?

1) There is 15 ml = 15 cm³ of water. It weighs 14.89 g.

2) Use the formula: density = weight ÷ volume

$$= 14.89 ÷ 15 = 0.99266...$$

The units are g/cm³ as the weight is in g and the volume is in cm³

$$= \mathbf{0.99\ g/cm^3}$$

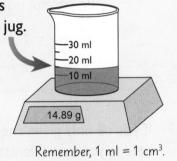

—30 ml
—20 ml
—10 ml

14.89 g

Remember, 1 ml = 1 cm³.

EXAMPLE 2:

Sarah cuts a piece of wood in the shape of a cuboid. It is 3 cm long, 2 cm wide and 1 cm thick. The wood has a density of 0.7 g/cm³.

What is the weight of the piece of wood that Sarah cut?

See page 93 for more on volume.

1) Work out the volume of the piece: volume = 3 × 2 × 1 = 6 cm³

2) Use the formula to find the weight: weight = density × volume
$$= 0.7 × 6 = \mathbf{4.2\ g}$$

Practice Questions

1) Diamond has a density of 3.5 g/cm³.
   What is the weight of a diamond that has a volume of 0.2 cm³?

   ....

2) A block of ice weighs 12 400 kg. It has the shape of a cylinder, with a radius of 1.2 m and a height of 3 m. What is the density of the ice? Give your answer to one decimal place.

   ....

   ....

Perimeter

Finding the Perimeter of a Shape

1) The perimeter is the distance around the outside of a shape.

2) To find a perimeter, you add up the lengths of all the sides.

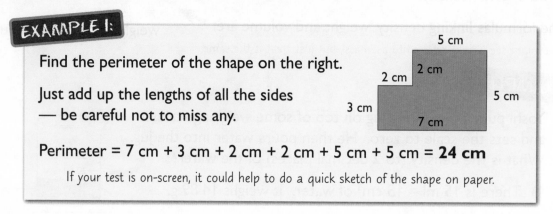

EXAMPLE 1:

Find the perimeter of the shape on the right.

Just add up the lengths of all the sides
— be careful not to miss any.

Perimeter = 7 cm + 3 cm + 2 cm + 2 cm + 5 cm + 5 cm = **24 cm**

If your test is on-screen, it could help to do a quick sketch of the shape on paper.

3) If you're only given the lengths of some of the sides, you'll have to work out the rest before you can calculate the perimeter. For squares and rectangles, this is fairly simple.

This square has four sides, but you're only given the length of one.

3 cm

For squares, all sides have the same length.

So the perimeter is 3 + 3 + 3 + 3 = 12 cm.

This rectangle has four sides, but you're only given the lengths of two of them.

9 cm

3 cm

For rectangles, sides that are opposite have the same length.

This is a '3 by 9' rectangle.

So the perimeter is 9 + 3 + 9 + 3 = 24 cm.

4) It's a bit harder to find the lengths of unknown sides if you're not dealing with rectangles.

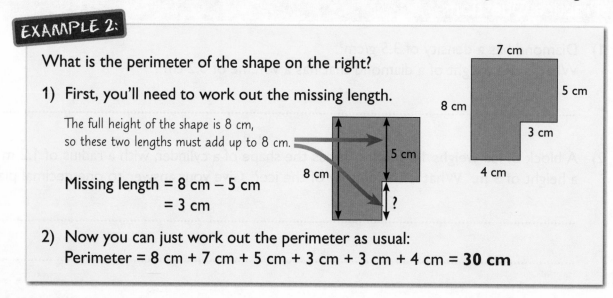

EXAMPLE 2:

What is the perimeter of the shape on the right?

1) First, you'll need to work out the missing length.

The full height of the shape is 8 cm,
so these two lengths must add up to 8 cm.

Missing length = 8 cm − 5 cm
　　　　　　　 = 3 cm

2) Now you can just work out the perimeter as usual:
Perimeter = 8 cm + 7 cm + 5 cm + 3 cm + 3 cm + 4 cm = **30 cm**

Changing a Perimeter by Removing Parts

If you remove a part of a shape, you can find the perimeter of the shape that's left over.

EXAMPLE:

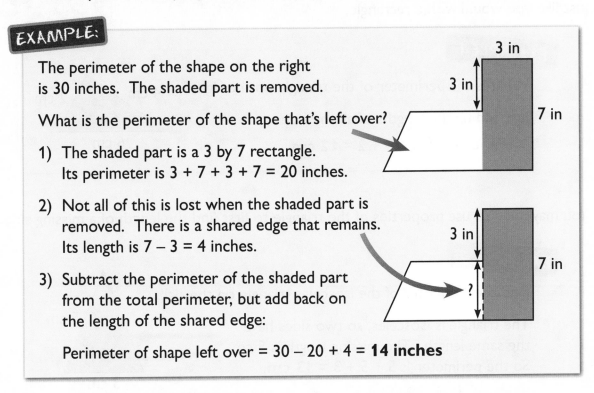

The perimeter of the shape on the right is 30 inches. The shaded part is removed.

What is the perimeter of the shape that's left over?

1) The shaded part is a 3 by 7 rectangle. Its perimeter is 3 + 7 + 3 + 7 = 20 inches.

2) Not all of this is lost when the shaded part is removed. There is a shared edge that remains. Its length is 7 – 3 = 4 inches.

3) Subtract the perimeter of the shaded part from the total perimeter, but add back on the length of the shared edge:

Perimeter of shape left over = 30 – 20 + 4 = **14 inches**

Practice Questions

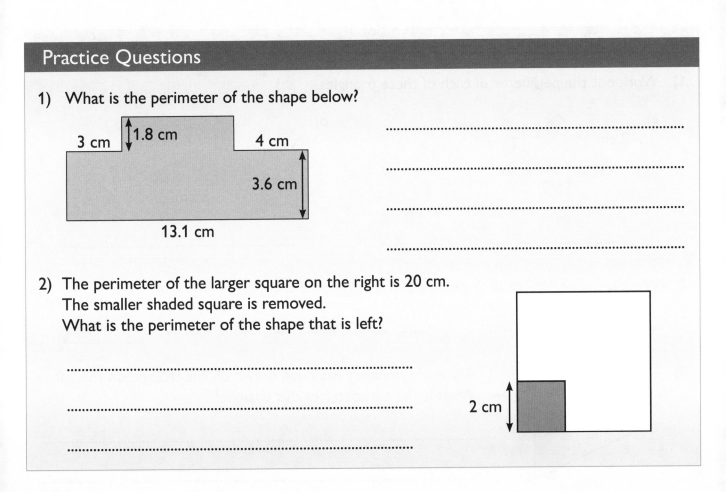

1) What is the perimeter of the shape below?

3 cm   1.8 cm   4 cm

3.6 cm

13.1 cm

....

....

....

....

2) The perimeter of the larger square on the right is 20 cm. The smaller shaded square is removed. What is the perimeter of the shape that is left?

2 cm

....

....

....

The Perimeter of a Triangle

You can find the perimeter of a triangle by adding up the lengths of all its sides — just like you would with a rectangle.

EXAMPLE 1:

What is the perimeter of the triangle on the right?

Just add up the lengths of all the sides:

Perimeter = 6 + 4 + 2 = **12 cm**

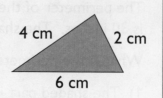
4 cm 2 cm
6 cm

You may need to use properties of the triangle to first find the length of a missing side.

EXAMPLE 2:

Find the perimeter of the isosceles triangle on the right.

The triangle is isosceles, so two sides have the same length. The missing length is 5 cm. So the perimeter is 5 + 5 + 3 = **13 cm**.

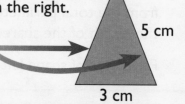
5 cm
3 cm

Practice Questions

1) Work out the perimeter of each of these triangles.

a)
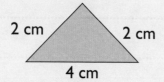
2 cm   2 cm
4 cm

b)
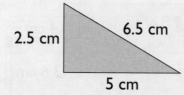
2.5 cm   6.5 cm
5 cm

....

2) What is the perimeter of the equilateral triangle on the right?

3 cm

....

3)
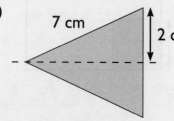
7 cm
2 cm

A line of symmetry has been drawn on the triangle on the left. What is the perimeter of this triangle?

....

....

Lengths in Circles Have Special Names

1) The diameter is the distance from one side of the circle to the other, passing through the centre.

2) The radius is the distance from the side of the circle to the centre. It's half the diameter.

3) The circumference is the perimeter of the circle — the distance all the way around the edge.

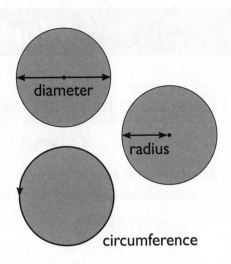

Finding the Circumference of a Circle

1) To work out the circumference of a circle, you have to use a special number called 'pi'. You write pi using the symbol π.

2) π is a decimal number that goes on forever. To three decimal places, it's 3.142.

3) If your calculator has a π button, you can use that just like any of the number keys. You might be told what value to use in place of π. (For example, 3.142, 3.14 or $\frac{22}{7}$).

4) The formula for calculating the circumference of a circle is:

Circumference of circle = π × diameter

You'll need to remember this formula. It won't be given in your test.

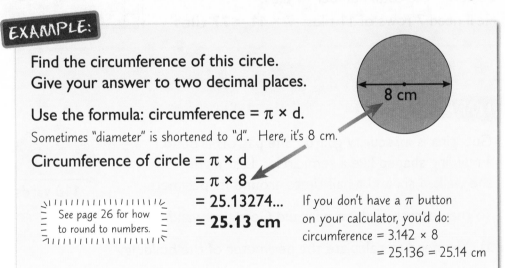

EXAMPLE:

Find the circumference of this circle.
Give your answer to two decimal places.

Use the formula: circumference = π × d.

Sometimes "diameter" is shortened to "d". Here, it's 8 cm.

Circumference of circle = π × d
= π × 8
= 25.13274...
= **25.13 cm**

See page 26 for how to round to numbers.

If you don't have a π button on your calculator, you'd do:
circumference = 3.142 × 8
= 25.136 = 25.14 cm

Be careful — if you're given the radius instead of the diameter, make sure to double it.

Practice Question

1) Find the circumference of each of these circles. Give your answers to two decimal places.

a) 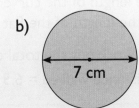 ....

....

b) ....

....

Working with Lengths

Questions Involving Length

1) There are lots of different types of question that involve length.

2) There's no single right way to answer them. Just use the information that you're given and work through it in a sensible way.

EXAMPLE 1:

Ruaridh is tiling his kitchen floor. The floor is 2.8 m wide and 4.4 m long. The tiles are 40 cm long and 40 cm wide. How many tiles will Ruaridh need?

1) Start by changing the sizes of the tiles from cm to m, so that all the lengths are in the same units:

    40 cm ÷ 100 = 0.4 m   Each tile is 0.4 m by 0.4 m.

2) Work out how many tiles he needs to make one row across the kitchen:

    4.4 m ÷ 0.4 m = 11 tiles

3) Work out how many rows are needed:

    2.8 m ÷ 0.4 m = 7 rows

4) Calculate the total number of tiles.
    He'll need 7 rows of 11 tiles: 7 × 11 = **77 tiles**

4.4 m

2.8 m

EXAMPLE 2:

Georgina is a security guard. She patrols around a building shaped like a semicircle. Last night, she walked six and a half times around its perimeter.

To the nearest yard, how far did Georgina walk?

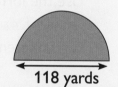

118 yards

A semicircle is a circle cut in half along its diameter.

1) You need to calculate the perimeter of the building.

    The length of the curved side of the building is half of the circumference of the full circle.

    → curved side = π × diameter ÷ 2
    = π × 118 ÷ 2
    = 185.353... yards

    The perimeter of the building is the length of the curved side, plus the length of the flat side.

    → perimeter = 185.353... + 118
    = 303.353... yards

2) Now work out the total distance Georgina has walked:
    distance walked = 6.5 × 303.353... = 1971.80... = **1972 yards**

Questions Involving Different Shapes and Lengths

Sometimes you'll have to deal with different shapes and lengths in one question.

EXAMPLE:

Alan wants to build a feature wall at the end of his patio. The two types of bricks he has and the pattern he wants to arrange them in are shown below. He needs to leave a 1 cm gap between each of the bricks. Gaps at the ends that aren't big enough for whole bricks will be filled with concrete.

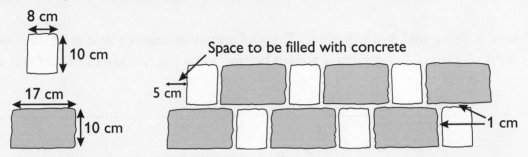

The wall needs to be 325 cm across and will be two rows high.
How many of each type of brick will he need?

1) Group the lengths together into a section of wall that will repeat over and over. For example...

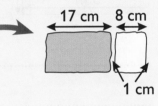

One large brick, a gap, one small brick and another gap has a total length of: 17 + 1 + 8 + 1 = 27 cm

2) The wall needs to be 325 cm across.

   325 ÷ 27 = 12.037... So the section of wall will repeat 12 times in one row.

   Work out the exact length of the wall if this section is repeated 12 times:

   12 × 27 = 324 cm   So there will be a 1 cm space at the end.
   No brick will fit there, so it will be filled with concrete.

3) The second row of the wall starts 5 cm further in than the first row.
   So the amount of space left to fill with bricks is 325 − 5 = 320 cm.

   320 ÷ 27 = 11.85...   So the section of wall will repeat 11 times.

   The exact length of the wall if this section is repeated 11 times is:

   11 × 27 = 297 cm   So there will be 320 − 297 = 23 cm left over.
   One more large brick could fit in this space.

4) Work out the total number of bricks.

   1st row: 12 large bricks and 12 small bricks

   2nd row: 12 large bricks and 11 small bricks

   Total: 24 large bricks, 23 small bricks

Practice Questions

1) Asif is wallpapering a wall in his dining room. The wall is 3.4 m wide. The wallpaper strips are 40 cm wide. How many strips will Asif need to buy to cover the width of the wall?

....

....

....

2) Jane is laying turf in the garden. The turf comes in squares that are 50 cm wide and 50 cm long. A plan of her garden is shown below. How many squares of turf will she need?

7.5 m

2.5 m

5 m   1.5 m

....

....

....

....

....

3) Sarah is setting up an exam hall with rows of desks facing the front of the hall. The hall is 12 m wide and 15 m long. Each desk is 70 cm wide and 50 cm long. She needs to leave a 1 m gap between each desk. Starting in a corner, how many desks can she fit into the hall?

....

....

....

....

Area

You Can Find the Area of Shapes by Multiplying

1) Area is how much surface a shape covers.

2) You can work out the area of squares and rectangles by multiplying the lengths of the sides together.

EXAMPLE 1:

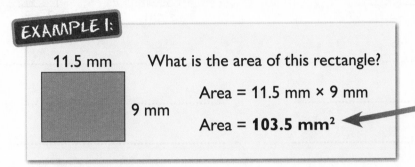

What is the area of this rectangle?

Area = 11.5 mm × 9 mm

Area = **103.5 mm²**

The units for the area are millimetres squared (mm²) because you've multiplied two lots of mm together.

EXAMPLE 2:

Lily's lawn has the shape of a square with a width of 41.3 ft. What is the area of the lawn?

Answer: 41.3 × 41.3 = **1705.69 sq. ft**

Imperial units of area are usually written as, for example, sq. ft (square feet) instead of ft².

Sometimes You Need to Split Shapes Up to Find the Area

You might have to find the area of an unfamiliar shape.

Do this by splitting it into shapes that you can find the area of.

EXAMPLE:

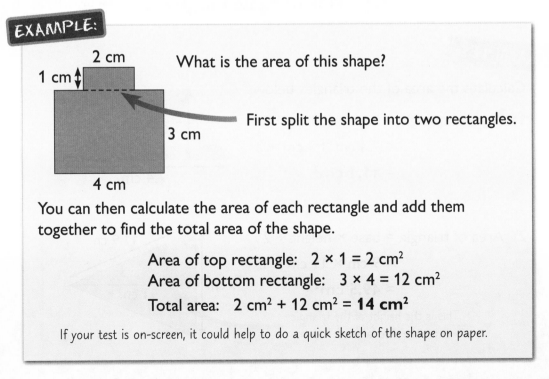

What is the area of this shape?

First split the shape into two rectangles.

You can then calculate the area of each rectangle and add them together to find the total area of the shape.

Area of top rectangle: 2 × 1 = 2 cm²
Area of bottom rectangle: 3 × 4 = 12 cm²
Total area: 2 cm² + 12 cm² = **14 cm²**

If your test is on-screen, it could help to do a quick sketch of the shape on paper.

Practice Question

1) Find the area of the shapes below.

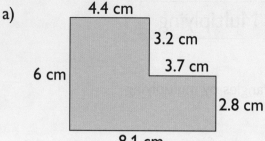

a) 4.4 cm, 3.2 cm, 3.7 cm, 6 cm, 2.8 cm, 8.1 cm

b) 13 cm, 11 cm, 11 cm, 5 cm, 3 cm, 3 cm

....

....

....

....

....

....

....

....

Learn the Formula for the Area of a Triangle

1) To work out the area of a triangle, you need to know its height and its base.

2) The height is the distance from the base to the opposite corner.

3) The formula for calculating the area is:

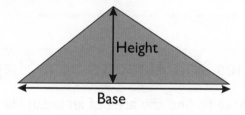

Area of triangle = base × height ÷ 2

You'll need to remember this formula. It won't be given in your test.

EXAMPLES:

Calculate the area of the triangles below.

1) Area of triangle = base × height ÷ 2
   = 7.4 cm × 3 cm ÷ 2
   = **11.1 cm²**

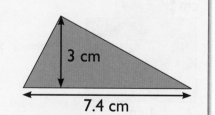

3 cm, 7.4 cm

2) Area of triangle = base × height ÷ 2
   = 9 cm × 11 cm ÷ 2
   = **49.5 cm²**

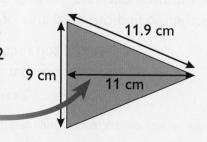

11.9 cm, 9 cm, 11 cm

This is the height of the triangle — not the side labelled 11.9 cm.

To Calculate the Area of a Circle You Need to Use Pi (π)

1) There's also a formula for working out the area of a circle.

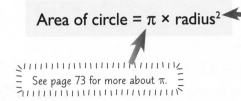

Area of circle = π × radius²

The little '2' means the radius is squared.
This means you multiply it by itself. So $r^2 = r \times r$.

See page 73 for more about π.

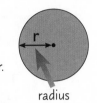

r

radius

2) You'll have to remember this formula — it won't be given in your test.

3) To work out the area, just put the radius of the circle into the formula.

EXAMPLE:

What is the area of the circle on the right?
Give your answer to two decimal places.

2.5 cm

Area of a circle = $\pi \times r^2$

$= \pi \times 2.5^2$

$= \pi \times 2.5 \times 2.5$

$= 19.63495...$

$= \mathbf{19.63 \ cm^2}$

Rounding is covered on page 26.

If you're given the diameter, you'll have to halve it to find the radius before using the formula.

If you don't have a π button on your calculator, you'd do:
area of a circle = 3.142 × 2.5 × 2.5 = 19.6375 = 19.64 cm²

Practice Questions

1) Find the area of the triangle below.

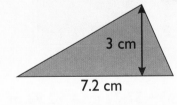

3 cm

7.2 cm

....

....

....

2) Find the area of the circle below to the nearest whole number.

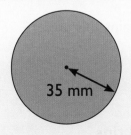

35 mm

....

....

....

Using Areas in Calculations

Sometimes you'll need to work out an area as part of a bigger calculation.

EXAMPLE 1:

Francine runs a slimming group.
The rectangular hall where the group meets is 8.6 m long and 7.2 m wide.
Health and Safety rules state for each person in the hall there must be 1.5 m²
of floor space. How many people (including Francine) can go to the group?

1) Work out the area of the hall. Area = 8.6 × 7.2 = 61.92 m²

2) Divide the area of the hall by 1.5 m² to find out how many
   people are allowed in the hall at a time.

$$61.92 ÷ 1.5 = 41.28$$

So **41 people** can go to the group.

EXAMPLE 2:

Joel is painting the side of a house, shown on the right.
A tin of paint will cover 20 m².
Joel needs to give the wall two coats of paint.

How many tins of paint does he need to buy?

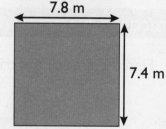

1) If you split the house into a triangle and a rectangle,
   you can work out the area of the side of the house.

*If your test is on-screen, it could help to do
a quick sketch of these shapes on paper.*

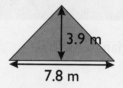
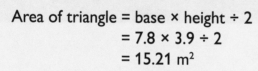

Area of triangle = base × height ÷ 2
   = 7.8 × 3.9 ÷ 2
   = 15.21 m²

Area of rectangle = 7.8 × 7.4
   = 57.72 m²

2) Then add the two areas together to find
   the total area of the side of the house: 15.21 m² + 57.72 m² = 72.93 m²

3) Joel needs to paint this area twice,
   so the total area that he will paint is: 72.93 × 2 = 145.86 m²

4) Now work out how many tins of paint Joel needs to paint this area.
   To do this, divide the total area he will paint
   by the area that one tin will cover:

$$145.86 \text{ m}^2 ÷ 20 \text{ m}^2 = 7.293... \text{ tins}$$

Joel can't buy 7.293 tins of paint, so he'll have to buy **8 tins**.

EXAMPLE 3:

Calculate the shaded area of this circle.
Give your answer to two decimal places.

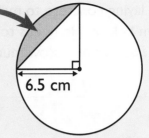

1) The diagram shows that the top-left quarter of the circle can be split into the shaded area and a right-angled triangle.

6.5 cm

Area of $\frac{1}{4}$ circle = $\frac{1}{4} \times \pi \times 6.5^2$

= 33.183... cm² ← To get the most accurate answer, use the full numbers and round only at the end.

Area of triangle = 6.5 × 6.5 ÷ 2

= 21.125 cm² ← Both the height and base of the triangle are equal to the radius of the circle.

2) To calculate the shaded area, subtract the area of the triangle from the area of the quarter circle.

Shaded area = 33.183... − 21.125 = 12.0581... = **12.06 cm²**

You Can Use Area Instead of Using Length

Some problems can be answered in lots of different ways. For example, you can work out the answer to some problems using length or using area.

EXAMPLE:

Ruaridh is tiling his kitchen floor. The floor is 2.8 m wide and 4.4 m long. The tiles are 40 cm long and 40 cm wide. How many tiles will Ruaridh need?

1) Start by changing the dimensions of the tiles from cm to m, so that all the lengths are in the same units:

40 cm ÷ 100 = 0.4 m   Each tile is 0.4 m by 0.4 m.

2) You need to work out how many tiles will fit into the area of the floor. So calculate the area of the floor: 2.8 × 4.4 = 12.32 m²

3) Then calculate the area of one tile: 0.4 × 0.4 = 0.16 m²

4) Now divide the area of the floor by the area of one tile:

12.32 ÷ 0.16 = **77 tiles**

This question was answered using lengths on page 74.

Practice Questions

1) Carlos is laying concrete to make the floors of two rooms. The dimensions of the rooms are shown below. He needs to use 0.1 m³ of concrete to make 1 m² of floor. The concrete will cost £65 per m³. How much will it cost to buy enough concrete for the floors?

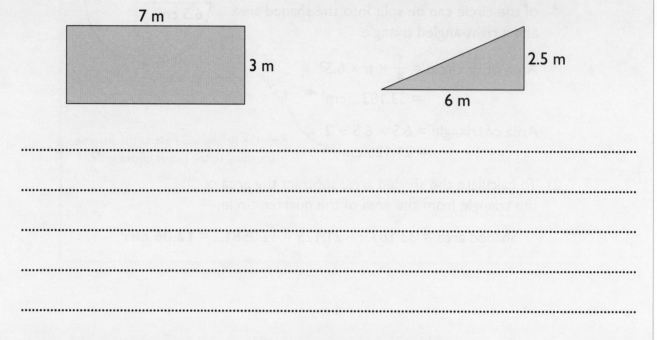

....

....

....

....

2) Samira is painting a ceiling. 1 L of paint can cover an area of 8 m².
A section of the ceiling is shown below. There are circular light fittings,
each with a diameter of 10 cm. Samira doesn't want to paint over these.

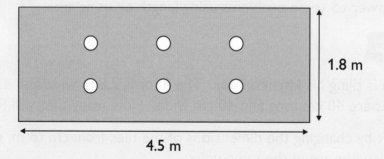

Show whether or not a 1 L can of paint is enough to paint this section of ceiling.

....

....

....

....

3D Shapes

Objects and Dimensions

1) Some objects are flat. Flat objects are called 2D objects.

2) Some objects are solid. Solid objects are called 3D objects.

3) The dimensions of an object tell you its size.

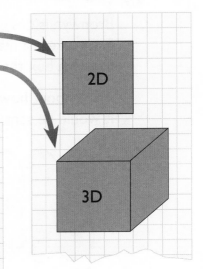

This box is 10 cm wide, 5 cm high and 6 cm deep.

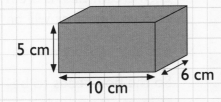

5 cm

10 cm

6 cm

The dimensions of the box are 10 cm by 5 cm by 6 cm.
This can also be written as 10 cm × 5 cm × 6 cm.

2D means '2-dimensional', so 2D objects have 2 dimensions — e.g. width and height.

3D means '3-dimensional', so 3D objects have 3 dimensions — e.g. width, height and depth.

You'll Need to Recognise These 3D Shapes

Cube   Cuboid

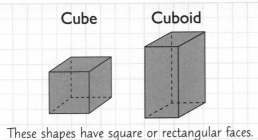

These shapes have square or rectangular faces.

Square-based Pyramid   Triangle-based Pyramid

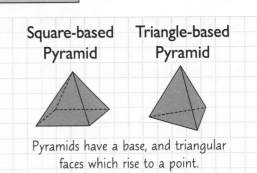

Pyramids have a base, and triangular faces which rise to a point.

Prisms

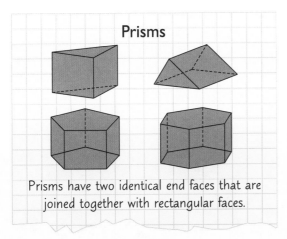

Prisms have two identical end faces that are joined together with rectangular faces.

Cylinder   Cone

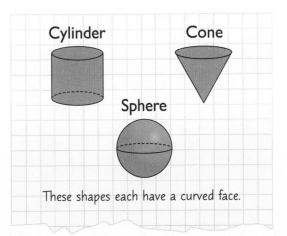

Sphere

These shapes each have a curved face.

Working with Dimensions

You might get a question about 3D objects and their dimensions in your assessment.

EXAMPLE:

Bronwyn is designing a box to hold 6 cream cakes.
Each cake is a maximum of 12 cm long, 5 cm wide and 4 cm high.
Sketch a box that could hold the cream cakes. Label the dimensions.

Think about how you'd put the cream cakes into a box.
It would be sensible to lie them all side by side in a row, like this:

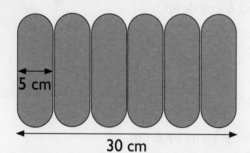

Each cream cake is 5 cm wide,
so 6 cream cakes will be:

6 × 5 cm = 30 cm wide.

So the box needs to be
at least 30 cm wide.

The cream cakes are all 12 cm long and 4 cm high.
So the box needs to be at least 12 cm long and 4 cm high.

So you could sketch a box
that looks like this:

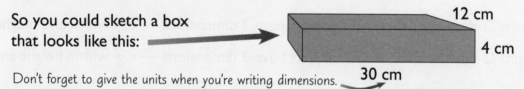

Don't forget to give the units when you're writing dimensions.

Practice Question

1) Stuart is posting some books to a friend. He wants a box to put them in.
The dimensions of the books are shown below.

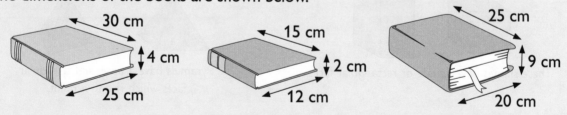

Sketch a box that could hold all the books. Label the dimensions of the box.

Nets

You Need to be Able to Draw Nets

A net is just a 3D shape folded out flat. You can use a net to help you make a 3D object.
The nets for cubes and cuboids always have the same basic shape.

EXAMPLE:

Draw a net for the cuboid on the right.

The cuboid has 6 faces, so a net for the cuboid will be made from 6 rectangles.

1) Draw the rectangle for the bottom of the cuboid first. It should be 11 cm long and 7 cm wide.

2) Next, draw rectangles for the sides of the cuboid. They're 11 cm long and 5 cm wide.

3) Now draw rectangles for the front and back of the cuboid. They're 5 cm by 7 cm.

4) Finally, draw the top of the cuboid with the same dimensions as the bottom.

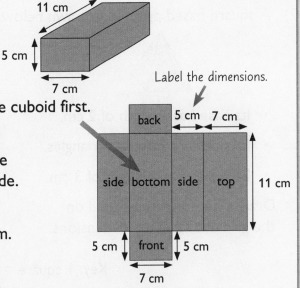

Label the dimensions.

Nets of Pyramids

The net of a pyramid will be made up of its base and all of its triangular sides.

EXAMPLE:

A triangle-based pyramid is drawn on the right.
The base is equilateral and the sides are isosceles.
Draw a net for this pyramid.

The pyramid has four faces, so its net should be made up of four triangles.

1) Draw the triangular base of the pyramid first. It's equilateral and the edges are 7 cm.

2) Now draw the three sides of the pyramid. Each side shares an edge with the base. These triangles are isosceles and the diagram shows that the two other edges are 9 cm long.

You should end up with something like this.

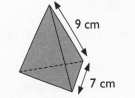

Practice Questions

1) Which two of the shapes on the right could be the net of a triangle-based pyramid?

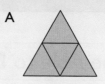
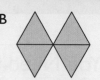

A   B

....

C   D

....

2) A square-based pyramid is drawn below.

- Its base has a width of 2 cm.

- Its sides are isosceles triangles.

- Each side has a height of 3 cm.

Draw a net for this pyramid on the grid and label the dimensions.

Key: 1 square = 1 cm

Nets of Prisms

The net of a prism is made up of rectangles and its two end faces.

EXAMPLE:

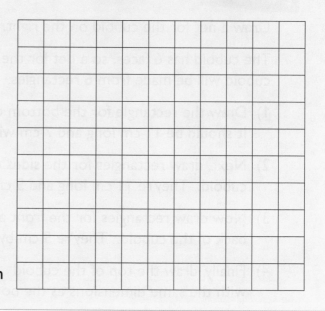

Draw a net of the triangular prism on the right.

The prism is made up of two triangles and three rectangles.

1) Draw the rectangles first. Use the diagram to work out their dimensions.

2) Then draw the triangles so that they each share an edge with one of the rectangles.

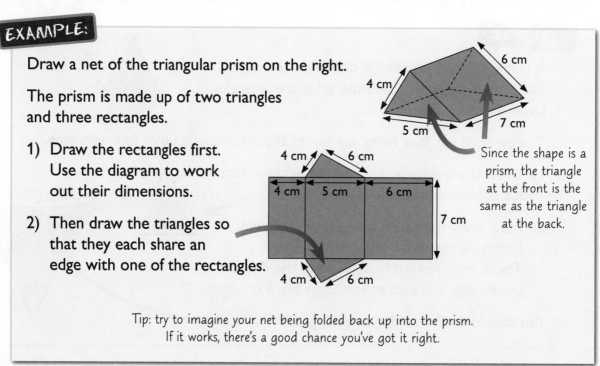

Since the shape is a prism, the triangle at the front is the same as the triangle at the back.

Tip: try to imagine your net being folded back up into the prism.
If it works, there's a good chance you've got it right.

Nets of Cylinders

A cylinder is a bit like a round prism. It has two circular faces joined by one curved side.

EXAMPLE:

Draw a net of the cylinder on the right.

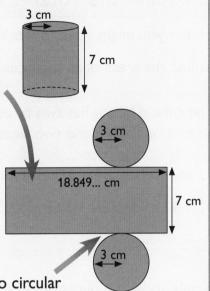

1) When you draw the net of a cylinder, the curved side flattens to a rectangle.

   The rectangle has the same height as the cylinder, and its width is equal to the circumference of the cylinder.

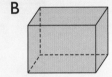

diameter = 2 × radius = 2 × 3 = 6 cm
circumference = π × diameter
= π × 6 cm
= 18.849... cm

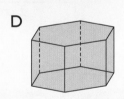

See page 73 for how to find the circumference of a circle.

2) Draw the rectangle, then draw the two circular faces anywhere along the edge of the rectangle.

Practice Questions

1) The net of a prism has been drawn below.

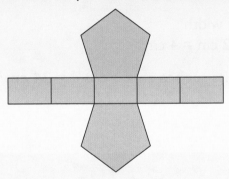

A   B   C   D

Which of the 3D objects above (A, B, C or D) does the net belong to? ....

2) A cylinder has been drawn below.

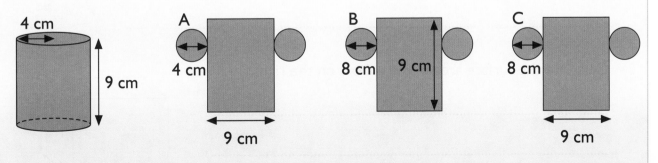

4 cm, 9 cm

A — 4 cm, 9 cm
B — 8 cm, 9 cm
C — 8 cm, 9 cm

Which diagram (A, B or C) is the correct net of the cylinder? ....

Surface Area

Surface Area is the Total Area of the Faces of a Shape

1) Sometimes you might need to work out the surface area of a shape.

2) This is just the areas of all the faces of the shape added together.

> The cuboid below has two faces with an area of 91 cm², two faces with an area of 70 cm², and two faces with an area of 32.5 cm².
>
> So the surface area of the cuboid is:
>
> 91 + 91 + 70 + 70 + 32.5 + 32.5 = 387 cm²

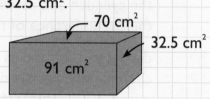
70 cm²
32.5 cm²
91 cm²

3) If you only know the dimensions of the faces, then you'll have to work out the areas.

EXAMPLE:

Calculate the surface area of this cube.

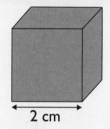

2 cm

1) The faces of a cube are identical squares, so they each have a length and width of 2 cm.

Area of face = length × width
= 2 cm × 2 cm = 4 cm²

2) The cube has six faces, so its surface area is 6 × 4 = **24 cm²**.

Practice Questions

1) The base of the pyramid on the right has an area of 25 cm².
The four other faces each have an area of 15 cm².
What is the surface area of the pyramid?

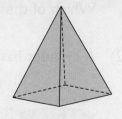

....

2) Calculate the surface area of the cuboid on the right.

....

....

....

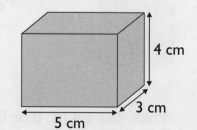
4 cm
3 cm
5 cm

Use a Net to Find the Surface Area of a Shape

If you need to find the surface area of a 3D shape, it can be helpful to draw a net.

EXAMPLE 1:

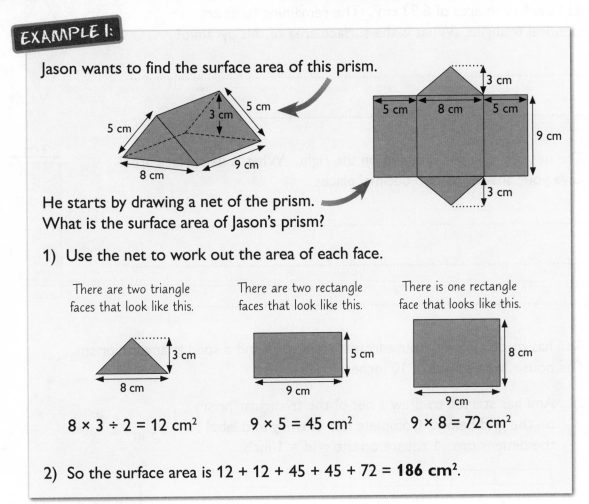

Jason wants to find the surface area of this prism.

He starts by drawing a net of the prism.
What is the surface area of Jason's prism?

1) Use the net to work out the area of each face.

There are two triangle faces that look like this.

There are two rectangle faces that look like this.

There is one rectangle face that looks like this.

$8 × 3 ÷ 2 = 12$ cm² $9 × 5 = 45$ cm² $9 × 8 = 72$ cm²

2) So the surface area is $12 + 12 + 45 + 45 + 72 = $ **186 cm²**.

EXAMPLE 2:

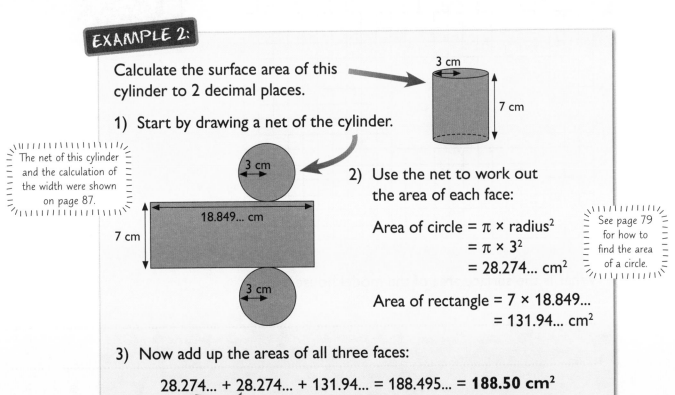

Calculate the surface area of this cylinder to 2 decimal places.

1) Start by drawing a net of the cylinder.

The net of this cylinder and the calculation of the width were shown on page 87.

2) Use the net to work out the area of each face:

Area of circle = $π × $ radius²
= $π × 3^2$
= $28.274...$ cm²

See page 79 for how to find the area of a circle.

Area of rectangle = $7 × 18.849...$
= $131.94...$ cm²

3) Now add up the areas of all three faces:

$28.274... + 28.274... + 131.94... = 188.495... = $ **188.50 cm²**

There are two circular faces, so don't forget to add this area twice.

Practice Questions

1) The net of a triangle-based pyramid is drawn on the right.
   Its base has an area of 6.93 cm². The remaining faces are
   identical triangles. What is the surface area of this pyramid?

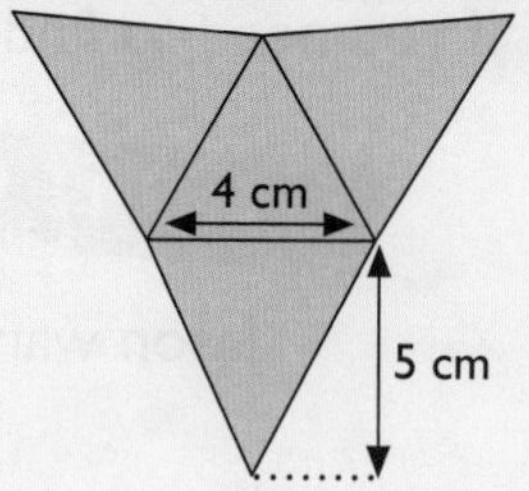

   ....

   ....

2) The net of a cylinder is drawn on the right. What is its surface area?
   Give your answer to two decimal places.

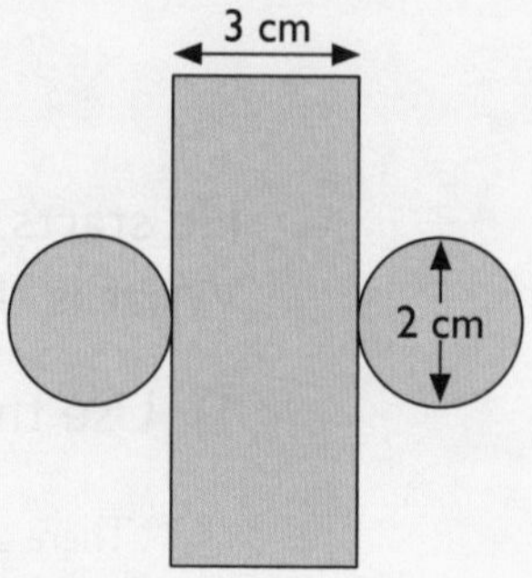

   ....

   ....

   ....

3) Ami has made a model house from a solid cube and a solid triangular prism.
   The house has a height of 10 inches.

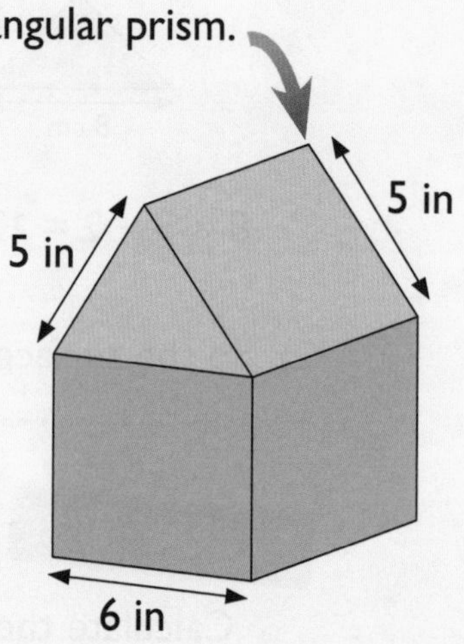

   a) Ami has started to draw a net of the triangular prism
      on the grid below. Complete her drawing and label
      the dimensions. 1 square on the grid = 1 inch.

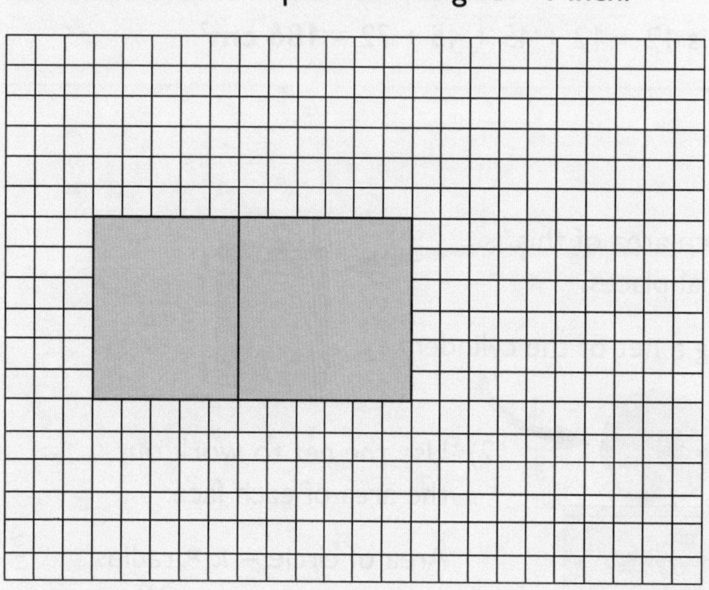

   b) What is the surface area of the model house?

   ....

   ....

   ....

Plans and Elevations

Plans and Elevations are 2D Drawings of 3D Shapes

1) You might be asked to draw accurate 2D drawings of 3D objects. How you draw them depends on where you're looking at the object from.

2) If you're looking at the object directly from above, the drawing is called a plan.

3) If you're looking at it from the front, it's called the front elevation (or front view).

4) If you're looking at it from the side, it's called the side elevation (or side view).

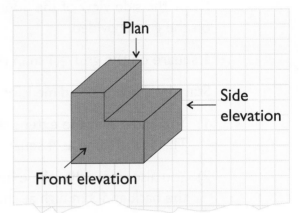

EXAMPLE 1:

Draw the plan, front elevation and side elevation of the shape on the right.

The shape has been made from identical cubes. Each cube has a side length of 1 cm.

1 square on the grid = 1 cm.

1) Start with the plan. If you look down on the object from above, you only see three of the cubes.

There are two cubes at the top and one sticking out from the side. The cube sticking out is lower down than the other two. Show the change in height by drawing a line on your plan.

2) From the front view, you see four cubes stacked one on top of another, and one cube sticking out from the side.

The cubes on the top and bottom are closer than the others, so draw two lines on your front elevation to show the change in depth.

3) From the side view, you can see all but one cube. One of the cubes is closer than the others. Show the change in depth by drawing two lines.

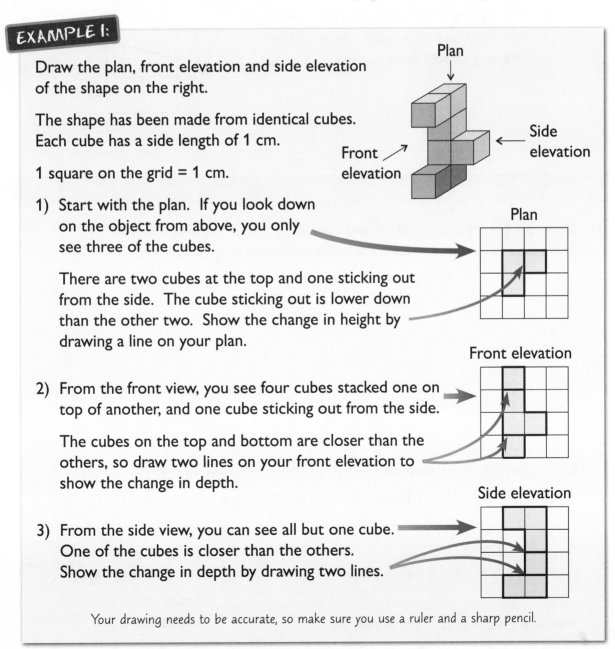

Your drawing needs to be accurate, so make sure you use a ruler and a sharp pencil.

For more complex shapes, you'll have to use the given dimensions when drawing plans or elevations.

EXAMPLE 2:

An architect is planning an extension to a house, shown on the right.

Draw the side view of the house on the grid provided.

1 square on the grid = 1 m in real life.

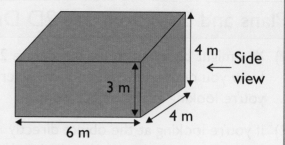

Start by drawing the dimensions you're given:

1) You know that the extension is 4 m wide. This is the same as 4 squares in your drawing.

2) You also know that the extension is 4 m high at the back and 3 m high at the front.

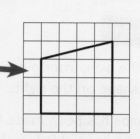

You can now draw the roof by connecting the front and back walls. Your finished drawing should look like this:

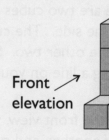

Practice Questions

1) Draw the plan, front elevation and side elevation of the shape below.
   The shape is made from identical cubes with side lengths of 1 cm.

   Plan   Front Elevation   Side Elevation

   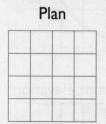   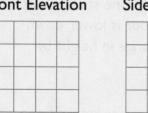

   1 square on the grid = 1 cm.

2) Jenine is camping outdoors. A sketch of her tent is shown below.
   Draw the plan, front elevation and side elevation of Jenine's tent.

   Plan   Front Elevation   Side Elevation

   1 square on the grid = 1 ft in real life.

   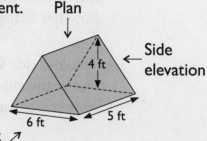

Volume

You Can Calculate the Volume of a Shape by Multiplying

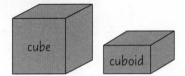

1) Volume is how much 3D space something takes up.

2) You can work out the volume of cubes and cuboids by multiplying the length, the width and the height together.

EXAMPLE:

Calculate the volume of the cube below.

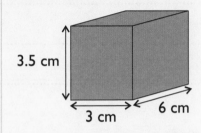

Length = 6 cm   Width = 3 cm   Height = 3.5 cm

Volume = length × width × height

= 6 cm × 3 cm × 3.5 cm

= **63 cm³**

3) The units are cm³ in this example because you've multiplied three lots of cm together. If the lengths were measured in m, the units for volume would be m³. Make sure that all the units are the same before multiplying the values together.

Using Volumes in Calculations

You may have to work out volume as part of a bigger calculation.

EXAMPLE:

Iain's fish tank is 0.9 m long, 0.3 m wide and 0.4 m high. He is filling it with water using a container that has a volume of 3000 cm³.

How many times will he need to empty the container into the tank to fill the tank?

1) Convert the sizes of Iain's fish tank into cm so that all measurements are in the same units:

0.9 m = 90 cm, 0.3 m = 30 cm and 0.4 m = 40 cm

2) Calculate the volume of the tank:

90 × 30 × 40 = 108 000 cm³

3) Calculate how many times the volume of the container will go into the volume of the tank:

108 000 ÷ 3000 = 36

So Iain will need to empty the container **36 times** to fill the tank.

Practice Questions

1) Calculate the volume of the shapes below.

a)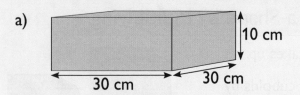

10 cm

30 cm   30 cm

b)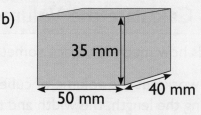

35 mm

50 mm   40 mm

....

....

....

....

....

....

....

....

2) Sarah has a suitcase that is 1.2 m long, 0.7 m wide and 20 cm deep.
What is the volume of the suitcase?

....

....

3) George is buying gravel to put on his driveway. The driveway is shown below.
He needs the gravel to be 2 cm deep. He can buy gravel in bags that cover 1 000 000 cm³.
How many bags will he need to buy?

12.2 m

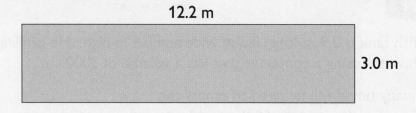

3.0 m

....

....

....

....

....

....

Finding the Volume of Prisms and Cylinders

1) A prism or a cylinder has two identical faces at either end of its length.
   You can call one of them the base.

2) Then you can calculate the volume of a prism or a cylinder using the formula:

   Volume = area of base × length

EXAMPLE 1:

Calculate the volume of this prism.

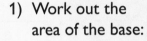

1) Work out the area of the base:

   Area of base = 6 × 4 ÷ 2
                = 12 m²

   The base of this prism is a triangle.

2) Multiply by the length:

   Volume = area of base × length
          = 12 × 7
          = **84 m³**

The base of a cylinder is a circle.

EXAMPLE 2:

Calculate the volume of this cylinder.

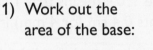

1) Work out the area of the base:

   Area of base = π × 2²
                = 12.566... m²

2) Multiply by the length:

   Volume = area of base × length
          = 12.566... × 6
          = **75.398... m³**

Practice Questions

1) Calculate the volume of the prism drawn below.

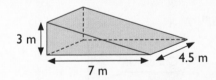

....

....

2) Find the volume of a cylinder (to one decimal place) that has...

   a) a radius of 4 cm and a height of 7 cm   b) a radius of 5 mm and a height of 12 mm

   ....

3) Simone drills a circular hole through a block of wood.
   The diameter of the hole is 1.5 inches.
   What is the volume of the wood that's left over?

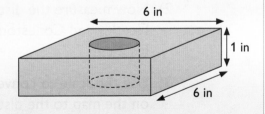

....

....

Scale Drawings

Scale Drawings Show Objects at Different Sizes

1) A scale drawing is a way of accurately drawing very big or very small objects on paper. All the dimensions are reduced or increased in the same way using the scale.

2) The scale is a ratio that tells you what a dimension in the drawing means in real life. For example, the scale 1:10 means that 1 cm in a drawing is equal to 10 cm in real life.

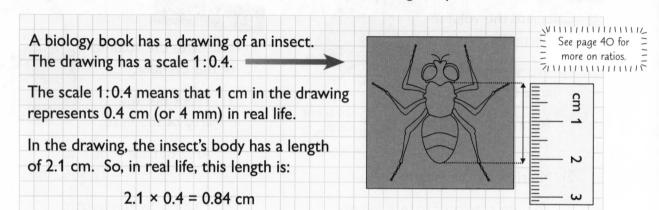

A biology book has a drawing of an insect. The drawing has a scale 1:0.4.

The scale 1:0.4 means that 1 cm in the drawing represents 0.4 cm (or 4 mm) in real life.

In the drawing, the insect's body has a length of 2.1 cm. So, in real life, this length is:

$$2.1 \times 0.4 = 0.84 \text{ cm}$$

See page 40 for more on ratios.

3) The scale could be given with units instead. For example, the scale 1:0.4 in the drawing could be written like "1 cm = 0.4 cm" or "1 cm = 4 mm".

4) You might also see the scale given as a line on the drawing. You'll need to use a ruler to measure the line to find what the real life dimension means.

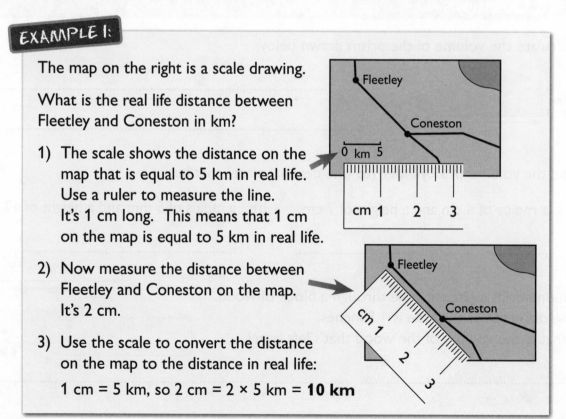

EXAMPLE 1:

The map on the right is a scale drawing.

What is the real life distance between Fleetley and Coneston in km?

1) The scale shows the distance on the map that is equal to 5 km in real life. Use a ruler to measure the line. It's 1 cm long. This means that 1 cm on the map is equal to 5 km in real life.

2) Now measure the distance between Fleetley and Coneston on the map. It's 2 cm.

3) Use the scale to convert the distance on the map to the distance in real life:
1 cm = 5 km, so 2 cm = 2 × 5 km = **10 km**

You may need to measure a distance in one unit and then convert it
to a more sensible unit after using the scale.

EXAMPLE 2:

Trevor is buying a new flat. The diagram shows
the layout of the rooms. It has a scale 1:150.

What is the width of the living room in metres?

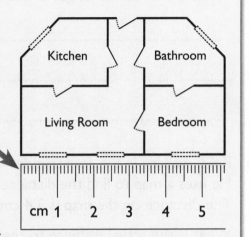

1) Use a ruler to measure the width of the
   living room on the diagram — it's 3.4 cm.

2) Use the scale to find the real life width.

   The scale is 1:150, so 1 cm on the diagram
   is 150 cm in real life, and 3.4 cm on the
   diagram is 150 × 3.4 cm = 510 cm in real life.

3) Finally, convert your answer from centimetres to metres.

   The living room has a width of 510 ÷ 100 = **5.1 m.**

Working Out the Scale

To find the scale used in a drawing, calculate the ratio
of the length in the drawing to the same length in real life.

EXAMPLE:

The map on the right is a scale drawing.
The real life distance between Fleetley
and Coneston is 10 km.

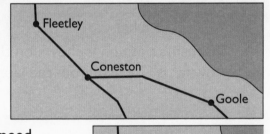

What is the scale used to draw this map?
Give your answer as a ratio.

1) You're told a real life distance, so you need
   to measure the same distance on the map:

   The distance on the map between
   Fleetley and Coneston is 2 cm.

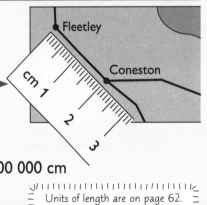

2) Both measurements must be in the same units,
   so convert the real life distance to cm:

   real life distance = 10 km = 10 000 m = 1 000 000 cm
   × 1000   × 100

   Units of length are on page 62.

3) Write the scale as a ratio and simplify:

   map distance : real life distance = 2:1 000 000
   = **1:500 000**

   See page 40 for how to simplify ratios.

   This means 1 cm on the map is equal to 500 000 cm (5 km) in real life.

Practice Questions

1) Drew has made a scale drawing of a bug.
   What is the real life length of the bug in inches?

   ....

   ....

   ....

   length

   $\frac{1}{5}$ inch

2) Kai is walking in the countryside.
   He uses a map to find the distance to the nearest village.
   The distance on the map is 3.4 cm. The map has the scale 1:100 000.

   What is the actual distance to the village in kilometres?

   ....

   ....

3) Shabnam is sailing on a lake. The diagram below is a scale drawing of the lake.

   Jetty

   Shabnam

   1.5 km

   a) The lake has a width in real life of 1.5 km.
      What is the scale used in the diagram as a ratio?

      ....

      ....

   b) Shabnam's position is shown on the diagram.
      How far is Shabnam from the jetty in metres?

   ....

   ....

4) Simon lives in Oaks. He travels by road to visit his friend Tariq in Cefn.
   How many miles does he travel?

   ....

   ....

   ....

   ....

   0   4
   miles

   Oaks

   Cefn   Furly

   A123

Making Scale Drawings

If you have to draw an object to scale, first work out what the dimensions should be.

EXAMPLE:

A rectangle has a height of 1.5 m and a width of 0.75 m. Draw the rectangle on the grid below using the scale 1:25. Each grid square is 1 cm².

1) The grid is in cm, so first convert the dimensions of the rectangle to cm:

real life height = 1.5 × 100 = 150 cm
real life width = 0.75 × 100 = 75 cm

2) The scale 1:25 means that 1 cm on the grid should represent 25 cm in real life.

The drawing will be smaller than the real thing, so you need to divide by 25 to get the drawing's dimensions:

drawing height = 150 ÷ 25 = 6 cm
drawing width = 75 ÷ 25 = 3 cm

Your scale drawing should look like this.

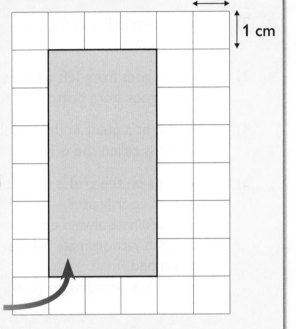

Practice Question

1) A scale drawing of Lucy's dining room is shown below.

   a) The real life dimensions of Lucy's fireplace are 200 cm by 50 cm.

   What is the scale used in the drawing as a ratio?

   ....

   ....

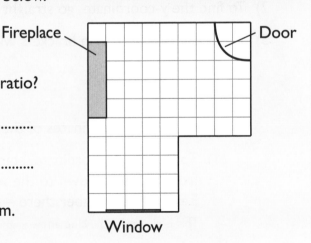

Lucy has bought a new table for her dining room. She wants to know where to put it.

- The table is 1.5 m wide and 2 m long.

- There should be at least 0.5 m all the way around the edge of the table.

   b) Using the correct scale, draw on the diagram where Lucy could put the table.

   ....

Coordinates

The Coordinate Grid

1) You can plot points and draw lines or shapes on a grid, like this one.

2) It's made by two lines crossing, called the axes.

3) The x-axis goes from left to right and the y-axis goes from bottom to top.

4) They meet at a point in the middle. This point is called the origin.

5) Every point on the grid is described by a pair of coordinates — (x, y). The x-coordinate always comes first and the y-coordinate always comes second.
The origin has coordinates (0, 0).

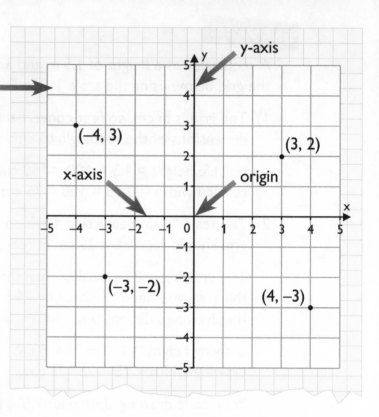

Reading Coordinates on a Grid

You need to be able to find the coordinates of a point marked on the grid.

1) To find the x-coordinate, go straight up or down from the point to the x-axis.

2) To find the y-coordinate, go straight left or right from the point to the y-axis.

3) Write the coordinates in brackets with the x-coordinate first and y-coordinate second.

EXAMPLE:

What are the coordinates of the point A on the grid below?

1) To find the x-coordinate, start from A and go down to the x-axis. Read off the number there — it's 3.
   This is shown by the blue arrow in the diagram.

2) To find the y-coordinate, start from A and go left to the y-axis. Read off the number there — it's 2.
   This is shown by the green arrow in the diagram.

3) So the point A has coordinates **(3, 2)**.

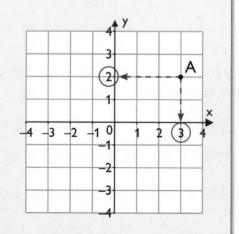

Plotting Points on a Grid

You can plot a point if you're given its coordinates:

1) Move left or right from the origin to find the x-coordinate on the x-axis.

2) Move up or down until you're in line with the y-coordinate on the y-axis.

3) Mark this position — this is your point.

EXAMPLE:

Plot the point (–2, 3) on this grid.

1) Move left from the origin to find –2 on the x-axis.

2) Move up until you're in line with 3 on the y-axis.

3) Mark this position.

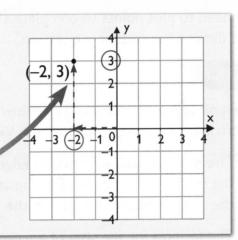

Practice Questions

1) Look at the grid on the right.

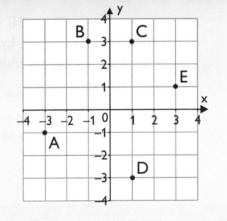

   a) Which of the points (A-E) is at (3, 1)?

   ....

   b) Which of the points (A-E) is at (–3, –1)?

   ....

2) Look at the grid on the right.
   Write down the coordinates of point...

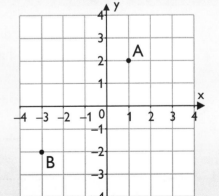

   a) A   b) B

   ....

   On the same grid, plot the following points:

   c) C = (–2, 3)   d) D = (3, –4)

Plotting Points to Make 2D Shapes

1) You can join up points on a grid to make 2D shapes.

2) For example, a square has been drawn on this grid by plotting four points (one for each corner) and joining them up with straight lines.

3) The square is named 'ABCD' after the letters used for its corners.

4) You may need to plot points to complete shapes that have dimensions given in a question.

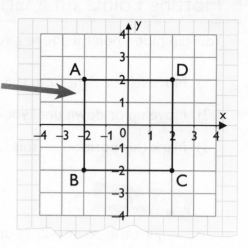

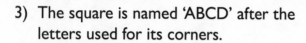
EXAMPLE:

The points A and B on the grid below are two corners of a triangle ABC. The triangle has an area of 10 cm². What could the coordinates of C be?

1) The triangle ABC will have an edge between A and B. Use this edge as the base of the triangle. Each square on the grid is 1 cm. By counting the squares between A and B, the length of this base is 4 cm.

2) The formula for the area of a triangle is:
   Area = base × height ÷ 2

   See page 78 for more on the area of a triangle.

   Write in the values you know:
   10 = 4 × height ÷ 2

   4 ÷ 2 = 2, so 10 = 2 × height.
   You multiply 2 by 5 to get 10,
   so the height must be 5 cm.

3) Count 5 squares up from the base. Any point with y-coordinate 3 will work. So point C could have coordinates **(0, 3)**.

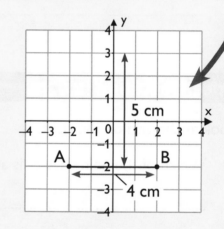

Key: 1 square = 1 cm²

Practice Question

1) On the grid, find the coordinates of points C and D such that the rectangle ABCD has a perimeter of 16 cm.

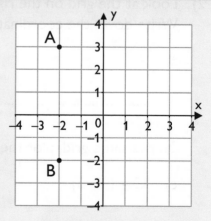

Key: 1 square = 1 cm²

....

....

....

....

Angles in 2D Shapes

Angles in a Triangle Add Up to 180°

1) If you add up the angles at the corners of a triangle, you'll always get the same answer.

> Angles in a triangle add up to 180°.

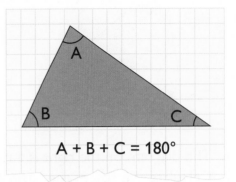

A + B + C = 180°

2) You can use this rule to work out the size of a missing angle when you know the other two.

EXAMPLE 1:

What is the size of the missing angle in this triangle?

The angles add up to 180°, so subtract the angles that you know from this:

Size of missing angle = 180° − 70° − 60° = **50°**

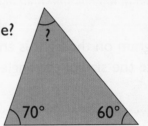

3) You might need to use other properties of triangles when finding the size of angles.

EXAMPLE 2:

The diagram on the right is an equilateral triangle. What is the size of the angle A?

1) You aren't told the size of any of the angles, but you know that the triangle is equilateral. So all three angles must be the same size.

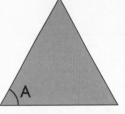

2) The three angles in the triangle add up to 180°, so divide by 3 to find the size of one angle. The size of angle A is 180° ÷ 3 = **60°**.

You can check your answer: 60° + 60° + 60° = 180°.

EXAMPLE 3:

An isosceles triangle has an angle of 100°. What are the sizes of the other two angles?

1) All three angles add up to 180°. So the other two angles add up to 180° − 100° = 80°.

2) An isosceles triangle has two equal angles, so the other two angles must both be 80° ÷ 2 = **40°**.

Practice Questions

1) Calculate the size of the missing angles in these triangles.

a)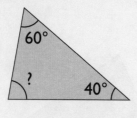

b)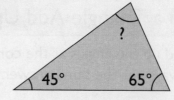

....

2) The diagram on the right is a right-angled triangle. Calculate the size of the angle A.

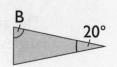

....

3) The diagram on the right is an isosceles triangle. Calculate the size of the angle B.

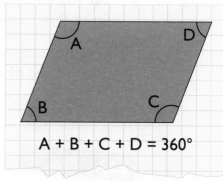

....

4) The diagram on the right shows a triangle with a line of symmetry. Calculate the size of the angle C.

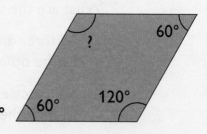

....

Angles in a Quadrilateral Add Up to 360°

A quadrilateral is a four-sided shape.

1) If you add up the angles at the corners of a quadrilateral, you'll always get the same answer.

> Angles in a quadrilateral add up to 360°.

2) You can use this rule to work out the size of a missing angle when you know the other three.

A + B + C + D = 360°

EXAMPLE 1:

What is the size of the missing angle in this shape?

The angles add up to 360°, so subtract the angles that you know from this:

Size of missing angle = 360° − 60° − 120° − 60 = **120°**

You might need to use other properties of quadrilaterals when finding the size of angles.

EXAMPLE 2:

The diagram below shows a quadrilateral with two lines of symmetry.
Calculate the size of the angles A, B and C.

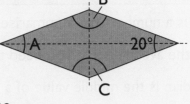

1) The lines of symmetry tell you that:
   - angles B and C are the same,
   - angle A is 20°.

2) To find the angles B and C, subtract the
   angles that you know from the total
   and divide the remaining angle by two. ➤

   $360° - 20° - 20° = 320°$
   $320° \div 2 = 160°$

So angle A is **20°** and angles B and C are each **160°**.

Practice Questions

1) Calculate the size of the missing angles in these quadrilaterals.

a)

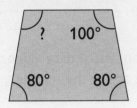

b)

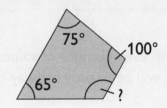

....

2) Josie has drawn the quadrilateral on the right.
   What is the size of the missing angle in her drawing?

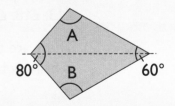

....

3) The diagram on the right shows a kite with one line of symmetry.
   Calculate the size of the angles A and B.

....

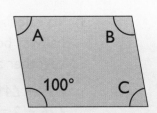

....

4) The diagram shows a parallelogram. It has two pairs of
   equal angles. Calculate the size of the angles A, B and C.

....

....

Median and Mode

The Median is a Type of Average

An average is a number that summarises a set of data.

There are two more averages you need to know about. The mode is on p.107 and the mean is on p.108.

1) One of the averages that you need to know about is called the median.

2) The median is the middle value of a set of data when the values are arranged in size order.

EXAMPLE 1:

Work out the median for the following data.

10, 6, 4, 7, 9, 2, 9, 3, 3, 7 and 9.

1) First, arrange the data in order of size: 2, 3, 3, 4, 6, 7, 7, 9, 9, 9, 10

2) The median is the middle value, which is **7**.

The easiest way to find this is to count in from each end of the arranged data until you have one number left.

3) If there's an even number of values, the median might not be an actual data value. Instead, it will be exactly halfway between the two values that are in the middle.

EXAMPLE 2:

A book shop records the amount of money that the last 10 customers spent:

£25, £35, £41, £60, £62, £18, £13, £28, £50, £100

What is the median amount of money spent?

1) Start by putting the data in order of size:
£13, £18, £25, £28, £35, £41, £50, £60, £62, £100

2) There are ten data values, so the middle of the data is halfway between the fifth and the sixth values.

£13　£18　£25　£28　£35　£41　£50　£60　£62　£100

3) To calculate the median, work out what number is exactly halfway between £35 and £41. Do this by adding them together and then dividing by 2:

£35 + £41 = £76
£76 ÷ 2 = **£38**

You might be able to work out what the halfway point is in your head. For example, the number that's halfway between 1 and 3 is 2.

The Mode is Also a Type of Average

The mode is the most common value that appears in a set of data.

EXAMPLE:

Work out the mode for the following data.

10, 6, 4, 7, 9, 2, 9, 3, 3, 7 and 9.

First, arrange the data in order of size: 2, 3, 3, 4, 6, 7, 7, 9, 9, 9, 10

The mode is **9** as it appears more than any other number (three times).

Practice Questions

1) Sharon has been timing her journey to work for the past 7 days. The times are: 45 mins, 36 mins, 29 mins, 40 mins, 32 mins, 38 mins, 44 mins. What is Sharon's median time?

....

....

2) Look at the numbers in the box on the right. What is the mode?

5	3	2	1	3
3	0	9	2	1

....

3) Jonathan makes boxes at a factory. The number of boxes he can make in 10-minute periods has been counted. The numbers he made were: 8, 7, 8, 7, 6, 8, 6.

a) What is the mode of the numbers of boxes that Jonathan made?

....

b) What is the median number of boxes that Jonathan made?

....

4) The weights of 6 cars are 1800 kg, 1750 kg, 1880 kg, 1940 kg, 1760 kg and 1820 kg.

a) What is the median weight?

....

....

b) Show a check of your answer to part a).

....

Mean and Range

The Mean is Another Type of Average

Another average that you need to know about is the mean.

Two other types of average (the median and the mode) were introduced on pages 106 and 107.

To work out the mean:

1) Add up all the numbers in the set of data.

2) Divide the total by how many numbers there are.

EXAMPLE:

The table shows the broadband speeds of a group of residents in a town. What is the mean broadband speed?

1) First, add up the numbers:
6.7 + 13.6 + 7.9 + 12.2 + 17.1 = 57.5

2) There are 5 numbers so divide the total by 5:
57.5 ÷ 5 = 11.5

3) The mean is **11.5 Mb per second**.

Resident	Speed (Mb per second)
Mr Stewart	6.7
Mrs Sharif	13.6
Mr Ward	7.9
Mrs Ford	12.2
Mrs Wells	17.1

The Range is the Gap Between Biggest and Smallest

The range is the difference between the biggest value and the smallest value.

To work out the range:

1) Write down all the numbers in order from the smallest to the biggest.

2) Subtract the smallest number from the biggest number.

EXAMPLE:

Babies are weighed when they are born. The weights of the babies born at a hospital this week in kilograms are: 2.2, 3.6, 2.6, 4.1, 4.0, 2.9, 2.4 and 3.2. Work out the range in the weights of newborn babies.

1) First, write the weights in order of size: 2.2, 2.4, 2.6, 2.9, 3.2, 3.6, 4.0, 4.1

2) Subtract the smallest number (2.2) from the biggest (4.1).
Range = 4.1 − 2.2 = **1.9 kilograms**.

Practice Questions

1) Calculate the mean and the range for each set of data below.

   a)   14   78   2   89   12

   ....

   ....

   b)   2.4   8.9   1.2   0.5   0   12.2

   ....

   ....

   c)   543   789   1003   24

   ....

   ....

2) Gilly stacks 12 boxes. The height of the stack is 36 feet.

   a)   Work out the mean height of the boxes.

   ....

   b)   The range of the heights of the boxes is 4 feet. The tallest box has
        a height of 5.2 feet. What is the height of the smallest box?

   ....

3) A vet weighs all the cats and all the dogs that she treats one day.

   a)   The mean weight of the cats was 3.9 kg. The total weight of all the cats was 42.9 kg.
        How many cats did she treat?

   ....

   b)   The mean weight of the dogs was 13.0 kg. She treated 8 dogs.
        The vet says that the range of the weights was 108 kg.
        Explain how you know that she's wrong.

   ....

   ....

   ....

Using Averages and Range

Averages and Range can be Used to Compare Sets of Data

You can use averages and the range in real life examples to compare data.

EXAMPLE 1:

Joe and Annette want to book a hotel. They have found some reviews online for two hotels (shown below). Each category has been scored out of 5.

Well Bridge Hotel		Review 1	Review 2	Review 3	Review 4	Review 5
Well	Location	4	4	3	3	3
Bridge	Service	4	3	3	3	4
Hotel	Rooms	4	3	2	3	3

		Review 1	Review 2	Review 3	Review 4	Review 5
Old	Location	5	4	5	4	4
Mill	Service	4	5	4	3	5
Hotel	Rooms	3	2	4	3	3

Which hotel has better reviews?

There's more than one way of answering this question, but you need to use the information in the table to support any answers you give.

For example, you could work out the mean score for location, service and rooms for each hotel and then compare them.

	Mean Score		
	Location	Service	Rooms
Well Bridge Hotel	3.4	3.4	3
Old Mill Hotel	4.4	4.2	3

You could use one of the other averages instead. For example, you could work out the medians and compare those.

From these means, the **Old Mill Hotel** looks like it has better reviews.

The range can be used to show how varied or consistent the data is.

If a set of data has a large range, then the greatest and smallest values are really far apart. This means the data is spread out and not very consistent.

EXAMPLE 2:

The table below shows the practice lap times set by three racing drivers. Which driver has been setting the most consistent times?

	Lap Times (seconds)				
Driver	Lap 1	Lap 2	Lap 3	Lap 4	Lap 5
Smithson	54	52	53	55	51
Olivier	49	56	55	57	53
Durango	51	54	56	55	52

1) To answer this you can use the range of times the driver has set.
A small range of lap times for a driver means their times are all quite similar, so they are consistent. A large range means that the driver's times are not very consistent.

Smithson's range = 55 secs − 51 secs = 4 seconds
Olivier's range = 57 secs − 49 secs = 8 seconds
Durango's range = 56 secs − 51 secs = 5 seconds

2) So the most consistent driver is **Smithson**, who had the smallest range of times (4 seconds) across his practice laps.

Practice Question

1) Jen wants to go on holiday with her husband.
The table below shows prices from four different travel companies.

	Fly Well	City Hols	Destination City	City Escapes
3 days	£580	£597	£479	£560
4 days	£635	£675	–	£730

a) What is the range and mean cost of a 3-day package from the four companies?

....

....

b) Destination City don't offer 4-day packages.
What is the range and mean cost of a 4-day package from the other three companies?

....

....

c) Jen thinks that Fly Well offers cheaper holidays than City Escapes. Is she right?

....

....

....

d) Are the prices more consistent for 3-day packages or 4-day packages?
Explain how you know.

....

....

You Can Use Averages to Make Estimates

You can use an average to make predictions.

For example, if you know a monthly average then you can use it to predict values for a whole year. To do this, you assume that every month has the average value, then multiply the average by the total number of months in a year (so you would multiply by 12).

> **EXAMPLE:**
>
> Heidi is late to work 3 times in November, 2 times in December, 8 times in January and 5 times in February. Work out the median and use this to find an estimate for the number of times that Heidi will be late in one year.
>
> 1) Work out the median by first putting the data in order: 2, 3, 5, 8. The median is halfway between 3 and 5, so it's 4.
>
> 2) Now assume that Heidi is late 4 times each month. So an estimate for the total number of times Heidi is late in a year is 4 × 12 = **48 times**.

Your estimate is unlikely to be exactly the same as the true value, but it should be fairly close.

In the example above, you assumed that Heidi will always be late 4 times a month. This probably won't be quite true (for example, she might be late more in winter because of the weather), but using the average should balance out the differences between the months.

Practice Question

1) Joan is the manager of a shop with 75 employees. She is looking at how long her employees have taken off work as holiday so far this year. She chooses 7 employees at random and writes down how many hours of holiday each of them has taken:

144   117   112   93   150   51   75

a) Use the mean number of hours of holiday for these employees to estimate the total number of hours of holiday taken so far this year by all of Joan's employees.

....

....

....

b) Each employee is allowed to take 150 hours of holiday. Use the median of the numbers above to estimate the total number of hours left to be taken by all of Joan's employees.

....

....

....

Grouped Frequency Tables

Grouped Data

Grouped frequency tables are like ordinary frequency tables, but they group data into classes.

Classes don't overlap.

In this table, no one has more than 11 pets. If you didn't know the greatest number, the last class should say something like '9 or more'.

Number of pets	Frequency
0 - 2	9
3 - 5	3
6 - 8	2
9 - 11	1

The frequency column tells you 'how many'. E.g. 9 people have between 0 and 2 pets.

Estimating the Mean for Grouped Data

1) When you have a grouped frequency table, you can work out an estimate of the mean of the data.

See p.108 for more on the mean.

2) Because the data is grouped, you don't know what the individual data values actually are. That's why the mean that you calculate is only an estimate.

To work out an estimate for the mean:

1) Add two extra columns to the table (these might already be drawn for you in the test).

2) Work out the midpoint of each class and write these in your first column.

For example, the midpoint of 0 - 2 is the number halfway between 0 and 2. That's 1.

Number of pets	Frequency	Midpoint	Frequency × Midpoint
0 - 2	9	1	9
3 - 5	3	4	12
6 - 8	2	7	14
9 - 11	1	10	10

3) Work out 'frequency × midpoint' and write these in your second column.

For the first class, frequency × midpoint = 9 × 1 = 9.

4) Work out the total frequency and the total 'frequency × midpoint' by adding up the columns.

Number of pets	Frequency	Midpoint	Frequency × Midpoint
0 - 2	9	1	9
3 - 5	3	4	12
6 - 8	2	7	14
9 - 11	1	10	10
Total:	15		45

5) The estimate for the mean is:

Mean = (Total 'frequency × midpoint') ÷ (Total frequency)

Mean = 45 ÷ 15 = 3 pets

Practice Questions

1) Stanley records the age of everyone in his snooker club in the table below.

Age	Frequency		
14 - 18	9		
19 - 23	3		
24 - 28	2		
29 - 33	1		
34 - 38	5		

Find an estimate for the mean age of the people in the snooker club.
You can use the empty columns in the table to help you.
You should give your answer to the nearest whole number.

....

....

....

2) The table below shows information about the number of calendars
sold on a calendar stall each month last year.

Number of calendars	Frequency
0 - 14	6
15 - 29	1
30 - 44	0
45 - 59	2
60 - 74	3
75 - 89	0

Work out an estimate for the average number of calendars sold each month.

....

....

....

Probability

Numbers can be Used to Describe Probability

Probability is a measure of how likely something is to happen.
Probabilities can be described using fractions, decimals or percentages.

- If something is **impossible**, it has a probability of 0 = 0%.

- If something has an **even chance** of happening, it has a probability of 0.5 = $\frac{1}{2}$ = 50%.

- If something is **certain**, it has a probability of 1.0 = 100%.

- If something is **likely**, the probability of it happening is between 0.5 and 1.
The more likely it is, the closer it will be to 1.

- If something is **unlikely**, the probability of it happening is between 0.5 and 0.
The less likely it is, the closer it will be to 0.

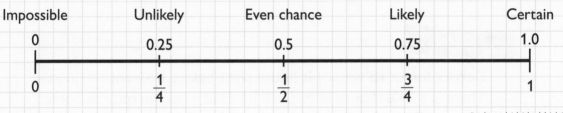

Impossible	Unlikely	Even chance	Likely	Certain
0	0.25	0.5	0.75	1.0
0	$\frac{1}{4}$	$\frac{1}{2}$	$\frac{3}{4}$	1

See p.35 for how to convert between fractions, decimals and percentages.

Calculating Probabilities

To calculate a probability, divide the number of ways that the
thing can happen by the total number of possible 'outcomes'.

$$\text{Probability} = \frac{\text{Number of ways for something to happen}}{\text{Total number of possible outcomes}}$$

An 'outcome' is just one thing that could happen.

EXAMPLE:

There are 250 boxes in a room. One of them contains a prize.
Robbie takes 195 of the boxes from the room.

What is the probability that the prize is still in the room?
Give your answer as a percentage.

Answer: There are 250 − 195 = 55 boxes left in the room.

Number of ways that the prize could still be in the room = 55
Number of possible outcomes = number of boxes = 250
Probability = $\frac{55}{250}$ As a percentage: 55 ÷ 250 × 100 = **22%**

The Probability Of Something Not Happening

If you know the probability that something will happen, you can work
out the probability that it won't happen by subtracting it from 1.

Probability that an event won't happen = 1 – Probability that the event will happen

If your probabilities are percentages, replace 1 with 100%.

EXAMPLE 1:

The probability that Norma's bus will be late is 0.6.
What is the probability that Norma's bus will **not** be late?

Answer: The event is 'Norma's bus is late'.
So the probability of the event happening is 0.6.

Probability that Norma's bus is not late
= 1 – Probability that it is late = 1 – 0.6 = **0.4**

EXAMPLE 2:

Derrick forgot to water his house plant before he went on holiday.
There is a 24% chance that it will be dead when he returns home.
What is the probability that it will still be alive?

Answer: 'Alive' means the same as 'not dead', so the probability of
the plant being alive is 100% – 24% = **76%**.

Multiply When an Event Happens More than Once

If you know the probability of an event happening, you can find the probability
of it happening lots of times. Just multiply the probability that you know
by itself that number of times.

EXAMPLE:

The probability of a weighted coin landing on heads is 0.9. It is tossed
three times. What is the probability that it lands on heads three times?

Answer: Multiply together three lots of 0.9
since the coin is tossed three times.

Probability of three heads = 0.9 × 0.9 × 0.9 = **0.729**

Practice Questions

1) Karen checks in for a flight and is given a seat at random.
There are four window seats, three aisle seats and nine centre seats available.
Calculate the probability that Karen is given a window seat. Write your answer as:

   a) a fraction

   ....

   b) a decimal

   ....

   c) a percentage

   ....

2) A garden centre has 1600 plant pots on sale. They cost £15 each.
Paul chooses some of the plant pots to buy at random, spending a total of £300.

   The garden centre realises that one of the plant pots that they had on sale was damaged.

   What is the probability that Paul bought the damaged plant pot?
   Give your answer as a decimal, correct to two decimal places.

   ....

   ....

3) A bag contains only green and orange balls. The probability of picking a green
ball out of the bag is 60%. What is the probability of picking an orange ball?

   ....

4) Jane has a weighted dice. When Jane rolls the dice, the probability that it lands on 6 is 0.3.

   a) What is the probability that the dice does **not** land on 6?

   ....

   b) Jane rolls the dice twice. What is the probability that it lands on 6 both times?

   ....

   c) Jane rolls the dice three times. What is the probability
   that it doesn't land on 6 on any of the rolls?

   ....

Probabilities from Diagrams

You might have to use a diagram to find the information you need. For example, the numbers you want might be in a table. You'll have to add up rows, columns or the whole table.

EXAMPLES:

A shop sells a selection of alarm clocks.
The table on the right shows some information about the alarm clocks that they sell.

	Radio	No radio
Analogue	5	6
Digital	3	11

The manager chooses an alarm clock at random to put in a window display.

1) What is the probability that the alarm clock is digital and has a radio?

   Answer: There are 5 + 3 + 6 + 11 = 25 alarm clocks in total.
   3 of these are digital and have a radio.
   So the probability is $\frac{3}{25}$ (or **0.12** or **12%**).

   There are 3 ways to pick this type of clock and 25 clocks to pick from (so 25 possible outcomes).

2) What is the probability that the alarm clock has no radio?

   Answer: Add up the 'no radio' column to find the total number without a radio: 6 + 11 = 17
   So the probability is $\frac{17}{25}$ (or **0.68** or **68%**).

Use a Diagram for Multiple Events

When there are two things happening (e.g. two spinners being spun),
you can use a special type of table to keep track of all of the results.

EXAMPLE:

The spinners on the right are spun and the outcomes added together. The table below shows the possible results. What is the probability that the result is 6?

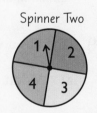

Spinner One   Spinner Two

If you roll a 2 on Spinner One and a 1 on Spinner Two, then the result is 2 + 1 = 3.

		Spinner One		
+	2	4	5	10
1	3	5	6	11
2	4	6	7	12
3	5	7	8	13
4	6	8	9	14

Spinner Two (rows 1, 2, 3, 4)

There are 16 possible outcomes of the two spins.

3 of these are '6'.

So the probability is $\frac{3}{16}$.

Practice Questions

1) Marwan is organising a party for the weekend. The table below shows how many of his guests are vegetarian and how many have an allergy. Marwan gives one of his guests a call.

	Vegetarian	Not vegetarian
Has an allergy	3	1
No allergies	10	7

a) What is the probability that the guest he calls is vegetarian and has an allergy?

....

b) What is the probability that the guest he calls is vegetarian?

....

2) The table on the right shows some information about the marital status of the people who filled in a survey.

	Married	Unmarried	Total
Aged 30 and under	562	159	721
Aged over 30		821	
Total	1213		2193

a) Complete the table.

b) Someone aged 30 or under is chosen at random. What is the probability that they are unmarried?

....

3) Percy has two special dice. The first dice is labelled 1, 3, 5, 6, 9, 10. The second dice is labelled 1, 2, 4, 6, 8, 10. Percy rolls both dice and then multiplies the results together.

a) Complete the table on the right to show the possible results.

b) What is the probability that the result of Percy's rolls is 1?

....

c) What is the probability that the result of Percy's rolls is 6?

....

d) What is the probability that the result of Percy's rolls is 95?

....

		first dice					
	×	1	3	5	6	9	10
second dice	1	1		5	6		10
	2	2	6		12	18	
	4	4	12	20		36	40
	6	6		30	36		60
	8	8	24	40		72	80
	10	10		50	60	90	

Scatter Diagrams

Scatter Diagrams Show Correlation

1) A scatter diagram shows how closely two things are related.

2) This relationship is known as 'correlation'. It can be positive or negative.

3) If two things are correlated, the points on the diagram will be close to a straight line.

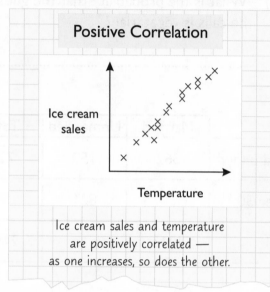

Positive Correlation

Ice cream sales

Temperature

Ice cream sales and temperature
are positively correlated —
as one increases, so does the other.

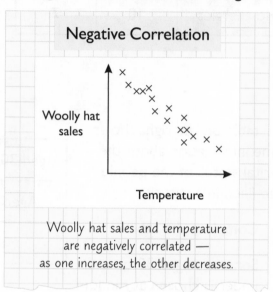

Negative Correlation

Woolly hat sales

Temperature

Woolly hat sales and temperature
are negatively correlated —
as one increases, the other decreases.

4) If there is no correlation between the two things, then the points will be scattered all over the diagram.

Use a Line of Best Fit to Predict Values

1) A line of best fit is a straight line that shows the correlation. It passes through roughly the middle of the set of points with about the same number of points above as below it.

2) If you know one value (e.g. the temperature in the examples above), you can read off the line to predict the other value (e.g. ice cream or woolly hat sales).

EXAMPLE:

12 people are looking for a job. The scatter diagram on the right shows the number of jobs that each of them applied for and the number of interviews they were invited to.

Use the line of best fit to predict how many interviews you would be invited to if you applied for 8 jobs.

1) Read up from 8 on the 'Jobs applied for' axis until you get to the line of best fit.

2) Read across to the 'Interviews invited to' axis.

So you can predict that you'd be invited to **4 interviews**.

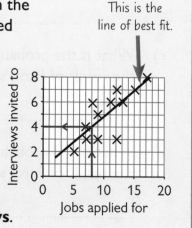

This is the line of best fit.

Interviews invited to

Jobs applied for

To Draw a Scatter Diagram, Plot the Data

EXAMPLE:

The table on the right shows the age and value of 8 cars. Draw a scatter diagram for this data.

Age (years)	Value (£)
1	7000
2.5	5500
5	2000
1.5	6000
3.5	3000
2	4500
4	2500
3.5	4500

1) Start by drawing your axes.
   One should show 'Age' and the other should show 'Value'.

   Make sure 'Age' goes up to at least 5 and 'Value' goes up to at least 7000.

Each gap on this axis represents £500.

Each gap on this axis represents half of a year (0.5 years).

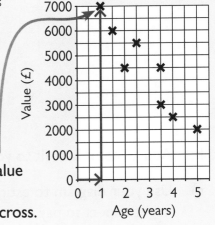

2) Then plot the data as points.

   For example, the car that is 1 year old has a value of £7000. So go across to 1 on the 'Age' axis, then up to £7000 on the 'Value' axis. Draw a cross.

Drawing a Line of Best Fit

To add a line of best fit to a scatter diagram, draw a straight line that follows the direction of the points. There should be roughly the same number of points on each side.

EXAMPLE:

Add a line of best fit to the scatter diagram that you drew above.

The line will be sloping downwards through the set of points. There should be about 4 on each side of the line.

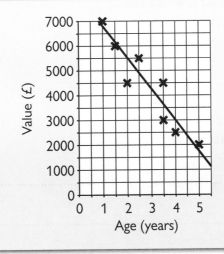

The line of best fit doesn't have to go through any of the points. There just needs to be roughly the same number on either side.

Practice Questions

1) The table below shows some information about 10 flats available to rent.

Number of rooms	2	4	5	8	5	4	2	8	9	7
Monthly rent (£)	300	450	500	750	600	400	350	900	950	800

   a) Draw a scatter diagram for this information on the grid below.

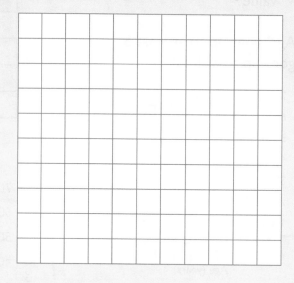

   b) Add a line of best fit to your scatter diagram.

   c) Use your diagram to estimate the monthly rent you
      might expect to pay for a flat with 6 rooms.

   ....

2) The scatter diagram on the right shows the
   average temperature and total monthly rainfall
   recorded at 15 weather stations last March.

   a) Use the diagram to predict the
      average temperature when the
      total monthly rainfall is 45 mm.

      ....

   b) Describe the relationship that is
      shown in the scatter diagram.

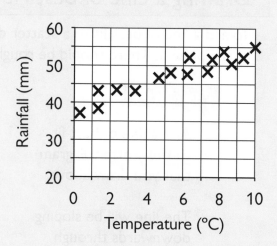

   ....

   ....

Candidate Surname		Candidate Forename(s)	

Test Date	Candidate Signature

Functional Skills

Mathematics   Level 2

Time allowed: 1 hour, 45 minutes

Section 1 — Non-calculator

You **may not** use a calculator.	

There are **15 marks** available for this section.

Section 2 — Calculator

You **may** use a calculator.	

There are **45 marks** available for this section.

You must have:
Pen, pencil, eraser, ruler, calculator (Section 2 only).

Instructions to candidates
- Use **black or blue ink** to write your answers.
- Write your name and the date in the spaces provided above.
- There are **2 sections** in this paper.
  Answer **all questions** in each section in the spaces provided.
- In calculations, show clearly how you worked out your answers.
- If your calculator does not have a π button, take the value of π to be 3.142.
- Check your working and answers.

Information for candidates
- Diagrams are not accurately drawn, unless otherwise stated.
- The marks available are given at the end of each question.

Section 1 — Non-calculator

Check your working as you go along.

You must not use a calculator in this section.

Q1.

Write 6 as a fraction of 18.
Give your answer in its simplest form.

$$\frac{\boxed{}}{\boxed{}}$$

(1 mark)

Q2.

Look at the numbers below.

| 36 | 39 | 13 | 48 | 27 | 24 | 63 |

What is the median?

_____

(1 mark)

Q3.

1 mile = 1.6 km

What is 15 miles in kilometres?

(tick one box)

A ☐ 9.375 km

B ☐ 16 km

C ☐ 20 km

D ☐ 24 km

(1 mark)

Q4.

Put these numbers in order from largest to smallest: 4904, 4094, 9400, 4090

_____   _____   _____   _____

(1 mark)

Q5.

$$\frac{20 - 6 \times 2}{2^2} =$$

(tick one box)

A ☐ 2

B ☐ 4

C ☐ 7

D ☐ 16

(1 mark)

Q6.

What are the coordinates of point A on the grid below?

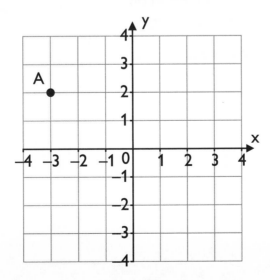

_____

(1 mark)

Q7.

27.6 ÷ 1.2 =

_____

(1 mark)

Q8.

What is $\frac{3}{4} - \frac{5}{8}$? Give your answer in its simplest form.

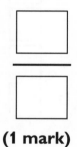

(1 mark)

Q9.

The points A, B, C and D have been plotted on the centimetre squared (cm²) grid on the right.

The shape ABCD is the face of a cube.

What is the cube's surface area?

(tick one box)

A ☐ 16 cm²

B ☐ 64 cm²

C ☐ 96 cm²

D ☐ 100 cm²

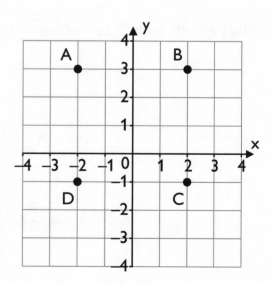

(1 mark)

Q10.

What is −12 + 26?

_____

(1 mark)

Q11.

Anita has a triangular flower bed in her garden. It has the following properties.

- The lengths of its sides are in the ratio 1:3:5.
- Its perimeter is 900 cm.

What is the length of the longest side of the flower bed?

(tick one box)

A ☐ 100 cm

B ☐ 500 cm

C ☐ 800 cm

D ☐ 900 cm

(1 mark)

Q12.

Keemia has a water bottle in the shape of a cylinder.
The base of the bottle has an area of 35 cm² and the height of the bottle is 20 cm.

What is the volume of her water bottle?

_____ cm³

(1 mark)

Q13.

Molly is mixing some paint.

She mixes 0.35 L of yellow paint with 0.2 L of blue paint.
Then she adds 0.015 L of white paint.

What is the total amount of paint that Molly has mixed?

_____ L

(1 mark)

Q14.

A catering company serves food at a party.

The time it takes them to serve all the guests is inversely proportional to the number of servers. Using 15 servers, all of the guests were served in 60 minutes.

How long would it have taken to serve all of the guests if 20 servers had been used instead?

_____ minutes

(1 mark)

Q15.

Riley is on a fishing trip for two weeks. After the first few days of the trip, he works out that the mean number of fish he has caught per day is 6.

Use this to work out an estimate for the total number of fish Riley will catch on his trip.

_____ fish

(1 mark)

End of Section 1

Section 2 — Calculator

Check your working as you go along.

You may use a calculator in this section.

Q1.

What is 24% of 62 500?

_____

(1 mark)

Q2.

Which of the following decimal numbers is the greatest?

(tick one box)

A ☐ 0.306

B ☐ 0.036

C ☐ 0.36

D ☐ 0.3006

(1 mark)

Q3.

The probability of an event happening is 0.77.

What is the probability that the event doesn't happen?

_____

(1 mark)

Q4.

The graph below can be used to convert between British pounds and Australian dollars.

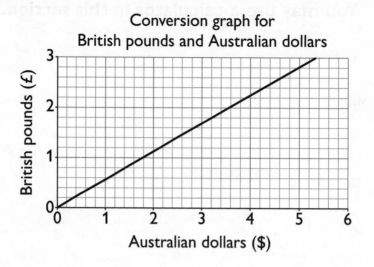

Use the graph to convert $3.20 into pounds.

£ _____

(1 mark)

Q5.

The shape below has one line of symmetry. It is not drawn accurately.

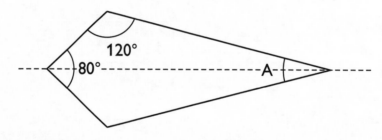

What is the size of angle A?

(tick one box)

A ☐ 40°

B ☐ 80°

C ☐ 120°

D ☐ 160°

(1 mark)

Q6.

Jane works at a leisure club. She asks a group of members how many hours of exercise they each do per week and also records their body mass index (BMI).

The scatter diagram shows this information.
A line of best fit has been drawn on for you.

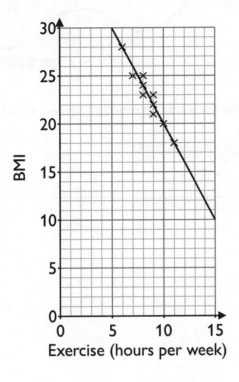

This formula is used to calculate body mass index: BMI = weight ÷ height^2

Jimmy's weight is 56 kg and he is 1.6 m tall. Using the line of best fit, Jane predicts that Jimmy does 7 hours of exercise per week.

Is her prediction correct?

Decision (yes / no) _____

Explanation and supporting calculations

(2 marks)

Q7.

Felix is competing in a cycling road race. The route for the race is sketched below.

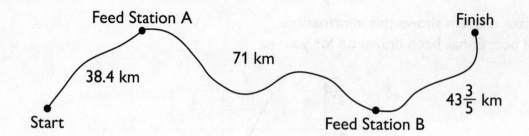

Feed Station A

Finish

38.4 km

71 km

$43\frac{3}{5}$ km

Start

Feed Station B

The race begins at 11:45. Felix cycles at an average speed of 32 km/h.

How far from the finish line is Felix at 16:15?

Show all your working.

_____ km

(5 marks)

Q8.

Natasha is organising an awards show. The diagram below shows her design for the trophy.

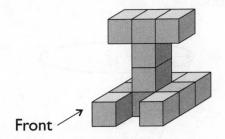

Front

The trophy is made of cubes. Each cube has a width of $1\frac{1}{2}$ inches.

Draw the front elevation of the trophy on the grid below.

1 in

1 in

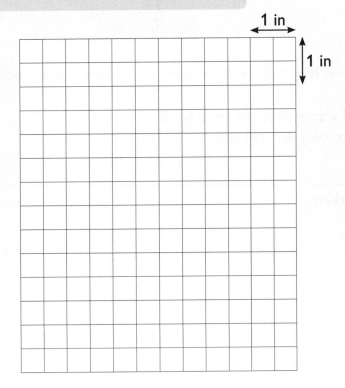

The base of the trophy is made of brass. The rest of the trophy is made of glass.
The brass part weighs 2.5 lb and the glass part weighs 1.0 lb.

What fraction of the total weight is made of glass?

(tick one box)

A ☐ $\frac{2}{5}$

B ☐ $\frac{5}{7}$

C ☐ $\frac{3}{5}$

D ☐ $\frac{2}{7}$

(3 marks)

134

Q9.

Tyler fills the cone below with water.
1 ml = 1 cm³ and 1 L = 1.76 pints.

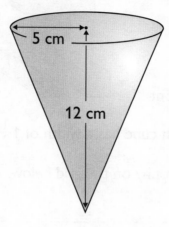

5 cm

12 cm

The volume of a cone is $\frac{1}{3}\pi r^2 h$, where r is the radius of the base and h is the height.

How many pints of water does the cone hold?
Give your answer to two decimal places.

Show all your working.

_____ pints

(3 marks)

Q10.

Takeshi takes out a loan of £195 000 to start a business.
He is charged 3.6% compound interest per year.

If Takeshi doesn't pay back any money, how much will he owe after two years?

Show all your working.

£ _____

(3 marks)

Q11.

Hattie is building a pond in her garden.
She wants to put a border made of pebbles around the pond.
The border will be 10 cm wide and is shown on the diagram by the shaded area.

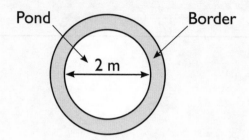

Pond   Border

2 m

Pebbles are sold in packs that cover 100 cm² and cost £4.80 each.

How much will it cost Hattie to buy the pebbles that she needs?

Show all your working.

£ _____

(5 marks)

Q12.

The table below shows the number of daily visitors to a nature reserve over several days.

Number of Visitors	Frequency		
0-14	4		
15-29	9		
30-44	14		
45-59	5		

Work out an estimate for the mean number of daily visitors.

Show all your working.

_____ **visitors**

(3 marks)

Q13.

The diagram below is a scale drawing of an area of land owned by the local council.
It is drawn accurately.

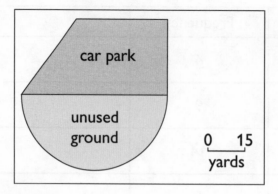

Calculate the area of the car park.

Show all your working.

_____ **square yards**

The unused ground next to the car park has the shape of a semicircle.
It is used to extend the car park.

Calculate the percentage increase in the total area of the car park.
Give your answer to the nearest whole number.

Show all your working.

_____ **%**

(8 marks)

Q14.

Dev has been counting the number of steps he takes each day.
The table below shows the data for one week. The data for Wednesday is missing.

Mon	Tue	Wed	Thu	Fri	Sat	Sun
5034	5412	?	5208	6117	3982	3472

Dev took the median number of steps on Wednesday.

What is the smallest number of steps that Dev could have taken on Wednesday?

Show all your working.

_____ **steps**

In the following week, Dev's daily step count had a range of 2071 steps.
Dev thinks his daily step count was more consistent during the second week.

Is Dev correct? Explain your answer.

Show all your working.

(4 marks)

Q15.

Madeline earns £1880 per month, before tax.
The following payments are taken from her pay each month.

Payment	Amount (per month)
National Insurance	12% on earnings above £719
Private pension	?
Income tax	£149

Madeline's take-home pay works out to be £1497.68 per month.

What percentage of her pre-tax pay is paid towards her private pension?

Show all your working.

_____ %

(4 marks)

End of Section 2

Answers — Practice Questions

Section One — Number

Page 4
Q1 Six million, nine hundred
and eighty-six thousand,
one hundred and ninety-seven
Q2 7 208 352
Q3 a) −16, −8, −2, 6, 13, 15
b) 14 568, 52 465, 574 643, 597 246
Q4 a) Rio   b) Vancouver
c) 14 °C
Q5 957 846, 957 824, 20 466,
−20 465, −571 468, −574 643,
−978 246
Q6 a) Joey   b) Oma

Page 6
Q1 14 034
Q2 15 960
check: 9660 + 15 960 = 25 620
Q3 £391
Q4 Yes, he has 80 hours left.

Page 8
Q1 4 787 317
Q2 5630 remainder 4
Q3 60
Q4 a) £1131.90   b) £3645.60

Page 9
Q1 121
Q2 51

Page 11
Q1 a) 61   b) 35
Q2 20
Q3 1
Q4 10

Page 14
Q1 $\frac{4}{13}$
Q2 45
Q3 7
Q4 a) $\frac{1}{6}$ b) $\frac{1}{2}$
c) $\frac{7}{9}$ d) $\frac{9}{2}$
Q5 a) $\frac{1}{5}$ b) $\frac{5}{8}$
c) $\frac{3}{4}$ d) $\frac{8}{5}$
Q6 $\frac{15}{17}$
Q7 $\frac{1}{5}$

Page 15
Q1 a) $\frac{13}{5}$ b) $\frac{10}{7}$
c) $\frac{41}{12}$ d) $\frac{28}{13}$
Q2 a) $2\frac{1}{3}$ b) $10\frac{3}{4}$
c) $4\frac{1}{5}$ d) $3\frac{3}{8}$

Page 16
Q1 $\frac{30}{36}$
Q2 a) $\frac{5}{20}, \frac{120}{400}, \frac{2}{5}$ b) $\frac{3}{20}, \frac{9}{54}, \frac{6}{24}$
Q3 Mike

Page 18
Q1 $\frac{13}{20}$
Q2 $\frac{1}{2}$
Q3 $\frac{44}{21}$
Q4 $\frac{61}{40}$
Q5 $4\frac{1}{2}$
Q6 $2\frac{3}{8}$
Q7 Yes, he has $21\frac{37}{40}$ kg of luggage.

Page 19
Q1 $\frac{29}{8}$
Q2 $\frac{21}{8}$
Q3 $\frac{1}{4}$

Page 20
Q1 9804
Q2 11 484
Q3 a) 315   b) 245

Page 22
Q1 Fifty-five point three five four
Q2 a) 62.48   b) 34.658
c) 46.83
Q3 0.0084, 0.185, 0.545
Q4 0.8 kg, 0.801 kg, 0.81 kg, 1.008 kg

Page 23
Q1 164.264
Q2 366.04
Q3 £238.32

Page 25
Q1 a) 5.568   b) 485.57
c) 10.27   d) 98
Q3 12.5 hours
Q4 £2.55

Page 27
Q1 a) 1.02   b) 3.6
Q2 a) 52   b) 5
Q3 70 seconds
Q4 243 cm²

Page 29
Q1 147
Q2 198
Q3 2507.2
Q4 210
Q5 1092

Page 30
Q1 26%
Q2 62.5%
Q3 37.5%
Q4 80%
Q5 55%
Q6 Brian. 75% of his customers have
sprinkles but only 67% (to the
nearest %) of Liv's customers have
sprinkles.

Page 31
Q1 £21 420
Q2 £36 750
Q3 45 552

Page 32
Q1 72
Q2 57 minutes
Q3 11 GB

Page 33
Q1 70 m²
Q2 125
Q3 £3600

Page 34
Q1 3% increase
Q2 18% increase
Q3 4% decrease

Page 36
Q1 a) 0.75   b) $\frac{11}{20}$
Q2 a) $\frac{5}{8}$ b) 0.6
Q3 0.9375

Page 38
Q1 30%
Q2 $\frac{23}{25}$
Q3 a) 0.59   b) 59%
Q4 12.5%

Page 39
Q1 a) $\frac{6}{15}$ b) 40%
c) 0.3   d) 67%
Q2 Deal 2. £7.60 is a bigger saving than
Deal 1, which saves £7.28

Page 42
Q1 300 ml
Q2 a) 48   b) 72
Q3 111 g
Q4 4 litres
Q5 320

Page 44
Q1 750 g
Q2 150 ml
Q3 88.2 minutes

Q4 No. For example, a ticket that costs £4 is for a distance of 16 km, but a ticket that costs three times more (£12) is not for a distance three times as far (the distance is four times as far).

Page 46
Q1 10 calls per hour
Q2 a) 24 minutes   b) 20 mph

Page 48
Q1 £35
Q2 £77.50
Q3 £137.50
Q4 £33
Q5 £68

Page 51
Q1 1176
Q2 3
Q3 4
Q4 a) $A = \frac{1}{2}h(a + b)$
    b) 54 cm²
Q5 19
Q6 £220

Section Two — Measures, Shape and Space

Page 52
Q1 a) 1642p   b) £2.10
Q2 11p or £0.11

Page 54
Q1 £95
Q2 £19.80
Q3 Car A

Page 55
Q1 £7.08
Q2 £113.40
Q3 No, the price should be £183.52.
Q4 50%
Q5 £2.43

Page 57
Q1 The 15-pack is best value for money.
    It's £0.52 per can. The price per can of the 6-pack is £0.53.
Q2 The free fitting offer will save Gillian the most money.
    It saves £120. The 20% off offer only saves £103.30.
Q3 The cheapest way for Josh to buy 12 bottles is to buy 3 packs of four.
    The pack of four offer is 60p per bottle. The other offer is 67.5p per bottle.

Page 58
Q1 £2050.57
Q2 £998.25

Page 59
Q1 a) £1200   b) 32%

Q2 £735

Page 60
Q1 £109.20
Q2 a) £4920   b) £21 320
    c) Yes, the new job pays more.
       There are 52 weeks in a year, so Ahmed's current job pays £21 320 per year.
Q3 3.25 hours or $3\frac{1}{4}$ hours or 3 hours and 15 minutes

Page 61
Q1 £22.35

Page 63
Q1 7500 m
Q2 6.4 kg
Q3 0.56 L

Page 64
Q1 33 lb
Q2 6.2 miles
Q3 a) 0.2 L   b) 200 ml
Q4 190.5 cm

Page 65
Q1 £375
Q2 375 ml
Q3 16 000 gallons

Page 66
Q1 a) 4 miles   b) 7 miles
Q2 a) 1 kg   b) 4 lb
    c) 4.4 lb

Page 67
Q1 8 m/s
Q2 10.4 mph

Page 68
Q1 a) The fastest route is A to B to C.
       It takes at least 2.1 hours. The road from A to C takes at least 2.2 hours.
    b) 75 miles

Page 69
Q1 0.7 g
Q2 913.7 kg/m³

Page 71
Q1 37 cm
Q2 20 cm

Page 72
Q1 a) 8 cm   b) 14 cm
Q2 9 cm
Q3 18 cm

Page 73
Q1 a) 37.70 cm   b) 21.99 cm

Page 76
Q1 9 strips
Q2 105 squares of turf
Q3 70 desks
    If you start with desks right against a side wall, you can fit 7 desks in a row

across the hall. If you start with desks right up against the front wall of the hall, you can fit ten rows going the length of the hall. 7 × 10 = 70 desks. If you left a gap at the front or side of the hall, you'll have a slightly different answer.

Page 78
Q1 a) 36.76 cm²
    b) 108 cm²

Page 79
Q1 10.8 cm²
Q2 3848 mm² (or 3849 mm² if you used 3.142 for π)

Page 82
Q1 £185.25
Q2 Yes, 1 L of paint is enough.
    The area of the ceiling is 8.1 m², but the six light fittings take up 6 × π × 0.1² = 0.188... m². So only 8.1 − 0.188... = 7.911... m² needs to be painted.

Page 84
Q1 For example:

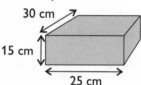

The easiest way to send the books is to stack them on top of each other. So the box needs to be as long and as wide as the biggest book and as high as all 3.

Page 86
Q1 A and D
Q2

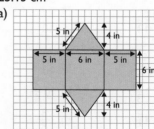

Page 87
Q1 C
Q2 C

Page 88
Q1 85 cm²
Q2 94 cm²

Page 90
Q1 36.93 cm²
Q2 25.13 cm²
Q3 a)

b) 264 cm²
You shouldn't include the face where the cube and prism meet because it's not on the surface of the combined shape.

Page 92
Q1

Plan   Front Elevation

Side Elevation

Q2
Plan   Front Elevation

Side Elevation

Page 94
Q1 a) 9000 cm³   b) 70 000 mm³
Q2 0.168 m³ or 168 000 cm³
Q3 1 bag

Page 95
Q1 47.25 m³
Q2 a) 351.9 cm³   b) 942.5 mm³
Q3 34.23... in³

Page 98
Q1 1 inch
Q2 3.4 km
Q3 a) 1 : 25 000   b) 500 m
Q4 34 miles

Page 99
Q1 a) 1 : 100
b) The table can go anywhere in the shaded area below. It should be 3 squares wide and 4 squares long.

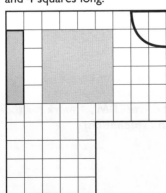

Page 101
Q1 a) E   b) A
Q2 a) (1, 2)   b) (−3, −2)
c) and d)

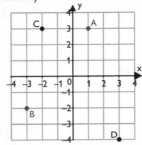

Page 102
Q1 C = (1, −2), D = (1, 3)

Page 104
Q1 a) 80°   b) 70°
Q2 65°
Q3 80°
Q4 100°

Page 105
Q1 a) 100°   b) 120°
Q2 40°
Q3 110°
Q4 B is 100°, A and C are 80°

**Section Three —
Handling Data**

Page 107
Q1 38 minutes
Q2 3
Q3 a) 8 boxes   b) 7 boxes
Q4 a) 1810 kg
b) For example, 1810 × 2 = 3620 and 1800 + 1820 = 3620.

Page 109
Q1 a) Mean = 39   Range = 87
b) Mean = 4.2   Range = 12.2
c) Mean = 589.75   Range = 979
Q2 a) 3 feet   b) 1.2 feet
Q3 a) 11 cats
b) If the mean is 13 and she treated 8 dogs, then the total weight of the dogs is 13 × 8 = 104 kg. The range can't be greater than the total, so the vet is wrong.

Page 111
Q1 a) Range = £118   Mean = £554
b) Range = £95   Mean = £680
c) Yes. On average, Fly Well is cheaper than City Escapes. The mean price of a Fly Well holiday (£607.50) is less than the mean price of a City Escapes holiday (£645).
d) The prices are more consistent for 4-day packages. You know this because the range is smaller.

Page 112
Q1 a) 7950 hours   b) 2850 hours

Page 114
Q1 24 years
Q2 30.75 calendars

Page 117
Q1 a) $\frac{1}{4}$ b) 0.25
c) 25%
Q2 0.01
Q3 40%
Q4 a) 0.7   b) 0.09
c) 0.343

Page 119
Q1 a) $\frac{3}{21}$ b) $\frac{13}{21}$
Q2 a)

	Married	Unmarried	Total
Aged 30 and under	562	159	721
Aged over 30	651	821	1472
Total	1213	980	2193

b) $\frac{159}{721}$
You could write this as decimal (0.22 to 2 d.p.) or a percentage (22% to the nearest whole number).

Q3 a)

first dice						
×	1	3	5	6	9	10
1	1	3	5	6	9	10
2	2	6	10	12	18	20
4	4	12	20	24	36	40
6	6	18	30	36	54	60
8	8	24	40	48	72	80
10	10	30	50	60	90	100

(second dice down the left side)

b) $\frac{1}{36}$ or 0.028 (3 d.p.) or 2.8%
c) $\frac{3}{36}$ or 0.083 (3 d.p.) or 8.3% (or an equivalent fraction, e.g. $\frac{1}{12}$)
d) 0

Page 122
Q1 a) and b)

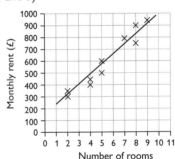

c) £650
Your answer will probably be slightly different. It will depend on how you've drawn your line of best fit.
Q2 a) 4 °C
Accept answers from 3 °C to 5 °C.
b) Positive correlation between temperature and rainfall (as temperature increases, so does rainfall)

Answers

Answers — Practice Paper

Section 1 — Non-calculator (Page 124)

1 $\frac{6}{18} = \frac{1}{3}$ *(1 mark)*

2 Put the numbers in order: 13, 24, 27, <u>36</u>, 39, 48, 63
   So the median is **36**. *(1 mark)*

3 10 miles = 1.6 × 10 = 16 km
   5 miles = 16 ÷ 2 = 8 km
   So 15 miles = 16 + 8 = **24 km**. *(1 mark)*

4 **9400, 4904, 4094, 4090** *(1 mark)*

5 $\frac{20 - 6 \times 2}{2^2} = \frac{20 - 12}{4} = \frac{8}{4} = 8 \div 4 = \mathbf{2}$ *(1 mark)*

6 **(−3, 2)** *(1 mark)*

7 $27.6 \div 1.2 = \frac{27.6}{1.2} = \frac{276}{12}$

$$12\overline{)2^2 7^3 6}$$
$$\,0\,2\,3$$

   So 27.6 ÷ 1.2 = **23** *(1 mark)*

8 $\frac{3}{4} - \frac{5}{8} = \frac{6}{8} - \frac{5}{8} = \frac{1}{8}$ *(1 mark)*

9 The side length of one face is 4 cm.
   So the area of one face is 4 × 4 = 16 cm².
   A cube has 6 faces, so the total surface area is 6 × 16.

   $$\begin{array}{r} 1\,6 \\ \times\ \ 6 \\ \hline 9\,6 \\ {\scriptstyle 3} \end{array}$$

   So the surface area of the cube is **96 cm²**. *(1 mark)*

10 **14** *(1 mark)*

11 1 + 3 + 5 = 9 parts
    9 parts = 900 cm, so 1 part = 900 ÷ 9 = 100 cm.
    The longest side is 5 parts = 100 × 5 = **500 cm**. *(1 mark)*

12 35 × 20 = 35 × 2 × 10 = 70 × 10 = **700 cm³** *(1 mark)*

13 Add up the three decimal numbers:

    $$\begin{array}{r} 0.3\,5\,0 \\ 0.2\,0\,0 \\ +\ 0.0\,1\,5 \\ \hline 0.5\,6\,5 \end{array}$$

    So Molly has mixed **0.565 L** of paint in total. *(1 mark)*

14 It would take 1 server 15 × 60 = 900 minutes.
    So it would take 20 servers
    900 ÷ 20 = **45 minutes**. *(1 mark)*

15 2 weeks = 2 × 7 = 14 days

    $$\begin{array}{r} 1\,4 \\ \times\ \ 6 \\ \hline 8\,4 \\ {\scriptstyle 2} \end{array}$$

    So you can predict that in 14 days he will catch
    a total of 14 × 6 = **84 fish**. *(1 mark)*

Section 2 — Calculator (Page 129)

1 62 500 × 0.24 = **15 000** *(1 mark)*

2 **0.36** *(1 mark)*

3 1 − 0.77 = **0.23** *(1 mark)*

4 Draw a line up from 3.2 on the $-axis to the line
   and then across to the £-axis:

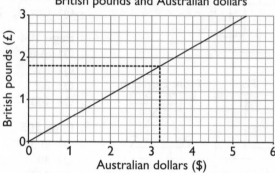

   Conversion graph for British pounds and Australian dollars

   So $3.20 is equal to **£1.80**. *(1 mark)*

5 The angle opposite the 120° angle is also 120°
   because the shape has one line of symmetry.
   So the sum of known angles is 80° + 120° + 120° = 320°.
   A = 360° − 320° = **40°** *(1 mark)*

6 **No** — Jimmy's BMI is 56 ÷ 1.6² = 21.875. Use the line
   of best fit to predict the number of hours of exercise
   that he does:

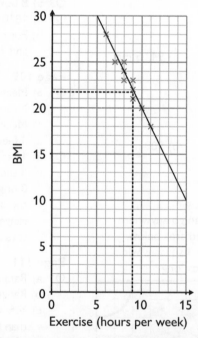

   So Jane should predict that Jimmy does 9 hours
   of exercise a week.
   ***(1 mark for Jimmy's correct BMI, 1 mark for the
   correct decision with explanation)***

7 $\frac{3}{5}$ = 0.6, so $43\frac{3}{5}$ km = 43.6 km

So the total length of the race is:
38.4 + 71 + 43.6 = 153 km.
11:45 to 16:15 is 4.5 hours, so Felix has been cycling for 4.5 hours.
Distance = speed × time = 32 × 4.5 = 144 km, so Felix has cycled 144 km. This means that he is 153 km − 144 km = **9 km** from the finish line.
(1 mark for converting all the distances into the same form, 1 mark for working out the total length of the race, 1 mark for working out the time, 1 mark for working out the total distance travelled, 1 mark for the correct final answer)

8

1 in

1 in

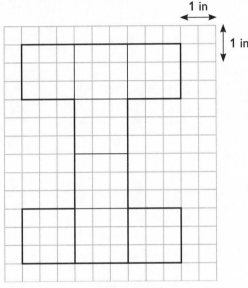

(1 mark for drawing eight 3 × 3 squares, 1 mark for correct front elevation)

Fraction made of glass = $\frac{1.0}{2.5 + 1.0} = \frac{1.0}{3.5}$

Simplify the fraction: $\frac{1.0}{3.5} = \frac{10}{35} = \frac{2}{7}$ *(1 mark)*

9 Volume of cone = $\frac{1}{3}\pi r^2 h = \frac{1}{3} \times \pi \times r^2 \times h$
$= \frac{1}{3} \times \pi \times 5^2 \times 12$
$= 314.15...$ cm³ *(1 mark)*
Amount of water = 314.15...ml
$= 0.31415...$ L *(1 mark)*
$= 0.31415... \times 1.76$
$= 0.55292...$ pints
$= \textbf{0.55 pints}$ (to 2 d.p.) *(1 mark)*

10 £195 000 × 0.036 = £7020
£195 000 + £7020 = £202 020
£202 020 × 0.036 = £7272.72
£202 020 + £7272.72 = £209 292.72
So Takeshi will owe **£209 292.72**.
(1 mark for using a correct decimal equivalent of 3.6%, 1 mark for a correct method of handling compound interest, 1 mark for the correct answer)
You'd get full marks if you used a slightly different method. For example, you could multiply £195 000 by 1.036 to get £202 020 in one step.

11 Radius of the pond = 2 ÷ 1 = 1 m = 100 cm
Area of the pond = $\pi \times r^2 = \pi \times 100^2$
$= 31\,415.9...$ cm² *(1 mark)*
Border width = 10 cm
Radius of the pond and border = 100 + 10 = 110 cm
Area of the pond and border = $\pi \times r^2 = \pi \times 110^2$
$= 38\,013.2...$ cm² *(1 mark)*
Border area = 38 013.2... − 31 415.9...
$= 6597.3...$ cm² *(1 mark)*
Number of bags needed = 6597.3... ÷ 100
$= 65.97...$ bags
So Hattie needs to buy 66 bags. *(1 mark)*
Cost of bags = 66 × £4.80 = **£316.80** *(1 mark)*

12

Number of Visitors	Frequency	Midpoint	Frequency × Midpoint
0-14	4	(0 + 14) ÷ 2 = 7	28
15-29	9	(15 + 29) ÷ 2 = 22	198
30-44	14	(30 + 44) ÷ 2 = 37	518
45-59	5	(45 + 59) ÷ 2 = 52	260
Total	32		1004

Mean = 1004 ÷ 32 = **31.375 visitors**
(1 mark for the correct midpoints, 1 mark for the correct total frequency × midpoints, 1 mark for correct estimated mean)
You could also round your final answer to the nearest whole number. This gives a mean of 31 visitors.

13 The measurements are:

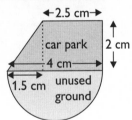

1 cm = 15 yards so 2 cm = 30 yards
2.5 cm = 37.5 yards
1.5 cm = 22.5 yards
Area of car park = (0.5 × 22.5 × 30) + (37.5 × 30)
$= 337.5 + 1125$
$= \textbf{1462.5 square yards}$
(1 mark for the correct measurements, 1 mark for correct conversions, 1 mark for a correct method for finding the area, 1 mark for the correct area)
Radius of semicircle in diagram = 4 ÷ 2 = 2 cm
Radius of real semicircle = 30 yards *(1 mark)*
Area of semicircle = $\frac{1}{2} \times \pi \times r^2$
$= \frac{1}{2} \times \pi \times 30^2$
$= 1413.7...$ *(1 mark)*
New car park area = 1462.5 + 1413.7...
$= 2876.2...$ square yards *(1 mark)*
New car park area as a percentage of the old area:
2876.2... ÷ 1462.5 = 1.9666...
$= 197\%$ (to the nearest %)
Percentage increase = 197% − 100%
$= \textbf{97\%}$ (to the nearest %) *(1 mark)*

14   There are 7 days, so the median is the 4th value.
     The given values in order are:
     3472, 3982, 5034, 5208, 5412, 6117
     So the 4th value lies between 5034 and 5208 *(1 mark)*.
     The smallest number of steps Dev could have
     taken on Wednesday is **5034 steps** *(1 mark)*.

     Range for 1st week = 6117 − 3472
                        = 2645 steps *(1 mark)*.
     2645 is greater than 2071 so Dev is **correct**.
     His daily step count was more consistent
     during the second week. *(1 mark)*

15   Earnings above £719 = £1880 − £719 = £1161 *(1 mark)*
     National Insurance = £1161 × 0.12 = £139.32 *(1 mark)*
     Private pension = £1880 − £1497.68 − £139.32 − £149
                     = £94 *(1 mark)*
     Percentage of pre-tax pay = 94 ÷ 1880
                               = 0.05 = **5%** *(1 mark)*

Glossary

12-hour Clock

The 12-hour clock goes from 12:00 am (midnight) to 11:59 am (one minute before noon), and then from 12:00 pm (noon) to 11:59 pm (one minute before midnight).

24-hour Clock

The 24-hour clock goes from 00:00 (midnight) to 23:59 (one minute before the next midnight).

2D Object

An object with 2 dimensions, i.e. a flat object.

3D Object

An object with 3 dimensions, i.e. a solid object.

Angle

A measurement of how far something has turned from a fixed point.

Anticlockwise

Movement in the opposite direction to the hands of a clock.

Area

How much surface a shape covers.

Average

A number that summarises a lot of data.

Axis

A line along the bottom and up the left-hand side of most graphs and charts. The plural is 'axes'.

BIDMAS

The correct order to carry out operations. It stands for Brackets, Indices, Division, Multiplication, Addition, Subtraction.

Capacity

How much something will hold. For example, a beaker with a capacity of 200 ml can hold 200 ml of liquid.

Certain

When something will definitely happen.

Circumference

The perimeter (distance around the outside) of a circle.

Clockwise

Movement in the same direction as the hands of a clock.

Coordinates

A pair of numbers used to describe the location of a point on a grid.

Correlation

A description of how closely two things are related. Correlation can be positive or negative, or there may be no correlation.

Data

Another word for information.

Decimal Number

A number with a decimal point (.) in it. For example, 0.75.

Density

A measure that relates the weight of something to its volume. Some common units of density are kg/m^3 and g/cm^3.

Diameter

The distance from one side of a circle to the other, going straight through the middle. The diameter is twice the radius.

Dimension

A number that tells you about the size of an object. For example, its length.

Estimate

A close guess at what an answer will be.

Even Chance

When something is as likely to happen as it is not to happen.

Formula

A rule for working out an amount.

Fraction

A way of showing parts of a whole. For example, ¼ (one quarter).

Frequency Table

A tally chart with an extra column that shows the total of each tally (the frequencies).

Front Elevation

A 2D diagram to show how a 3D object looks from the front.

Impossible

When there's no chance at all of something happening.

Interest

Money paid or added on to a value over time, given as a percentage.

Length

How long something is. Length can be measured in different units, for example, millimetres (mm), centimetres (cm), or metres (m).

Likely

When something isn't certain, but there's a high chance it will happen.

Line of Symmetry

A shape with a line of symmetry has two halves that are mirror images of each other. If the shape is folded along this line, the two sides will fold exactly together.

Mean

A type of average. To calculate the mean, you add up all the numbers and divide the total by how many numbers there are.

Median

A type of average. The median is the middle value of a set of data when the values are arranged in size order.

Mixed Number

When you have a whole number and a fraction together. For example, 2¼ (two and a quarter).

Mode

A type of average. The mode is the most common value that appears in a set of data.

Negative Number

A number less than zero. For example, −2.

Net

A 3D shape folded out flat. You can use a net to help you make a 3D object. For example, you can use a net to make a box.

Percentage

A way of showing how many parts you have out of 100. For example, twenty percent (20%) is the same as 20 parts out of 100.

Perimeter

The distance around the outside of a shape.

Pi (π)

π is a decimal number that goes on forever. To 3 decimal places, it's 3.142. It's used in the calculation of a circle's circumference and area.

Plan View

A 2D diagram to show how a 3D object looks from above.

Probability

The likelihood (or chance) of an event happening or not.

Profit

The difference between the cost of making something and the price it's sold for.

Proportion

A way of showing how much of one part there is compared to the whole thing. For example, if there are 4 towels and 1 of them is white then the proportion of white towels is 1 in 4.

Quadrilateral

A 2D shape with 4 straight sides and 4 corners.

Radius

The distance from the side of a circle to the middle. The radius is half the diameter.

Range

The difference between the biggest and smallest numbers in a data set.

Ratio

A way of showing how many things of one type there are compared to another. For example, if there are 3 red towels to every 1 white towel then the ratio of red to white towels is 3:1.

Right Angle

A square corner.

Scale

A rule that tells you how far a given distance on a drawing is in real life.

Scatter Diagram

A diagram of points plotted on axes that shows how two things are related.

Side Elevation

A 2D diagram to show how a 3D object looks from the side.

Speed

How fast something is moving. Some common units of speed are miles per hour (mph), kilometres per hour (km/h) and metres per second (m/s).

Square Number

A number multiplied by itself. For example, 5 squared (5^2) is the same as 5×5.

Symmetry

See line of symmetry.

Table

A way of showing data. In a table, data is arranged into columns and rows.

Triangle

A 2D shape with 3 straight sides and 3 corners.

Unit

A way of showing what type of number you've got. For example, metres (m) or grams (g).

Unlikely

When something isn't impossible, but it probably won't happen.

Volume

The amount of space something takes up.

Weight

How heavy something is. Grams (g) and kilograms (kg) are common units for weight.

Index

M2CGSRA1